Arthur Sherburne Hardy

Elements of the differential and integral calculus

Method of rates

Arthur Sherburne Hardy

Elements of the differential and integral calculus
Method of rates

ISBN/EAN: 9783742892379

Manufactured in Europe, USA, Canada, Australia, Japa

Cover: Foto ©berggeist007 / pixelio.de

Manufactured and distributed by brebook publishing software
(www.brebook.com)

Arthur Sherburne Hardy

Elements of the differential and integral calculus

ELEMENTS

OF THE

DIFFERENTIAL AND INTEGRAL CALCULUS.

METHOD OF RATES.

BY

ARTHUR SHERBURNE HARDY, Ph.D.,

Professor of Mathematics in Dartmouth College.

—∘∘⦂⊙⦂∘∘—

BOSTON, U.S.A.:

PUBLISHED BY GINN & COMPANY.

1890.

PREFACE.

THIS text-book is based on the method of rates, which, in the experience of the author, has proved most satisfactory in a first presentation of the object and scope of the Calculus. No comparisons have been made between this method and those of limits or of infinitesimals. This larger view of the Calculus, and of mathematical reasoning and processes in general, cannot readily be given with good results in the brief time allotted the subject in the general college course.

The immediate object of the Differential Calculus is the measurement and comparison of rates of change when the change is not uniform. Whether a quantity is or is not changing uniformly, however, the rate at any instant is determined in essentially the same manner; viz. by ascertaining what its change would have been in a unit of time had its rate remained what it was at the instant in question. It is this change which the Calculus enables us to determine, however complicated the law of variation may be. This conception of the nature of the problem is simple, and seems to afford the best foundation for further and more comprehensive study; while for those who are not to make a

special study of mathematics it secures a more intelligent
and less mechanical grasp of the problems involved than
other methods whose conceptions and logic are not easily
mastered in undergraduate ·courses.

My thanks are due to Professor Worthen, my colleague,
for valuable suggestions and assistance in the reading of
proofs.

ARTHUR SHERBURNE HARDY.

HANOVER, N.H., June 2, 1890.

CONTENTS.

—◦◦—

PART I. — THE DIFFERENTIAL CALCULUS.

CHAPTER I. — INTRODUCTORY THEOREMS.

CHAPTER II. — DIFFERENTIATION OF EXPLICIT FUNCTIONS.

The Algebraic Functions.

The Transcendental Functions.

The Logarithmic and Exponential Functions.

The Trigonometric Functions.

The Circular Functions.

CHAPTER III. — SUCCESSIVE DIFFERENTIATION.

Applications of Successive Differentiation.

Accelerations.

Development of Continuous Functions.

Evaluation of Illusory Forms.

Maxima and Minima.

CHAPTER IV. — FUNCTIONS OF TWO OR MORE VARIABLES.

CHAPTER V. — PLANE CURVES.

Curvature.

Evolutes and Envelopes.

CONTENTS. ix

PART II. — THE INTEGRAL CALCULUS.

CHAPTER VI. — TYPE INTEGRABLE FORMS.

CHAPTER VII. — GENERAL METHODS OF REDUCTION.

By Partial Fractions.

By Rationalization.

Part I.

THE DIFFERENTIAL CALCULUS.

CHAPTER I.

INTRODUCTORY THEOREMS.

1. Quantities of the Calculus. The quantities of the Calculus are, like those of Analytic Geometry :

Variables : whose values change continuously within the limits assigned by their mutual relations. Thus, in the equation of the circle $x^2 + y^2 = R^2$, x and y are variables having any and all values between the limits $\pm R$.

Arbitrary constants : as R in the above equation, which may have any arbitrarily assigned values, but which do not change when the variables change.

Absolute constants : which admit of no change whatever; such as R would become if the radius of the circle were assumed to be 5.

2. Functions. As in Analytic Geometry, also, any quantity is said to be a **function** of another when it depends upon the latter for its value. Thus, $a - x$, $\tan x$, $(a^2 - x^2)^{\frac{1}{2}}$, are functions of x. The variable upon which the function depends is called the **independent variable**.

An equation between two variables may be solved for either regarded as the function, the other being the independent variable. Thus, from $x^2 + y^2 = R^2$ we have $y = \sqrt{R^2 - x^2}$, in which y is the function and x the independent variable, or $x = \sqrt{R^2 - y^2}$, in which x is the function and y the independent variable. The distinction implies no difference in the nature of the variables, for each is dependent upon the other, and serves only to distinguish the variable whose values are assigned from that whose values are derived.

3

A quantity may depend upon several variables for its value, and is then said to be a function of two or more variables. Thus, $x^2 + y^2 - R^2$, xzv, are functions of two and three variables, respectively. If no condition is imposed upon the function, the variables are said to be independent. If, however, we subject the function to some condition, as $x^2 + y^2 - R^2 = 0$, x and y are said to be dependent, since they can only vary in such a way as to make the function zero. Although dependent upon, that is, functions of, each other, a value may be assigned to one and that of the other derived from the equation; either one may therefore be regarded as the independent variable in the sense explained above.

When the variables are dependent and their mutual relations are known, the function may be expressed in terms of any one regarded as the independent variable. Thus, the function xzv represents the volume of a parallelopiped whose edges are x, z, and v, and the variables are independent. If, however, we impose the conditions $x = mz$, $z = nv$, that is, if the ratios of homologous sides are to remain constant, the variables become dependent, and the function may be expressed in terms of any one, as $\dfrac{x^3}{m^2 n}$.

The conditions of the problem will determine whether the variables are dependent or independent, and in the former case the manner of their dependence.

3. Classification of functions.

I. Functions are classified as **algebraic** and **transcendental**. Algebraic functions are those which involve only the six fundamental operations of Algebra: addition, subtraction, multiplication, division, involution, and evolution, the indices in the latter cases being constant. All other functions are transcendental; the more common of which are:

The *logarithmic* function, $x = \log y$, and its inverse form, $y = a^z$, the *exponential* function;

The *trigonometric* functions, $y = \sin x$, $y = \cos x$, etc., and their inverse forms, $x = \sin^{-1} y$, $x = \cos^{-1} y$, etc., the *circular* functions.

An algebraic function of a single variable which contains no power of the variable above the first is called a *linear* function. Such can always be reduced to the form $mx + b$.

II. If an equation between several variables be solved for any one, the latter is said to be an **explicit** function of the others, the manner of its dependence being exhibited by the solution of the equation. Otherwise it is said to be an **implicit** function. Thus, in $x^2 + y^2 = R^2$, x and y are implicit functions of each other; while, in $y = \sqrt{R^2 - x^2}$, y is an explicit function of x. The difference is one of form only, the chief object of Algebra being the reduction of functions from implicit to explicit forms. The notation $y = f(x)$, $y = f'(x)$, $y = \phi(x)$, etc., read 'y a function of x,' is used to denote that y is an explicit function of x; and the notation $f(x, y) = 0$, $\phi(x, y) = 0$, etc., to denote that x and y are implicit functions of each other.

III. If in any function $y = f(x)$, y increases and decreases with x, y is called an **increasing** function of x; but if y decreases when x increases, or increases when x decreases, y is said to be a **decreasing** function of x.

The increase and decrease referred to is *algebraic*.

Thus, in $y = mx + b$, y is an increasing function of x; but in $y = -mx + b$, y is a decreasing function of x. Again, in $y^2 = 2px$, y has two values, one of which is an increasing, the other a decreasing, function of x.

If we plot the locus of $y = f(x)$, this relation of the variables to each other is represented graphically. Thus, $x^2 = y$ is a parabola situated as in the figure, from which we see that when x is negative, that is in the second angle, x is a decreasing function of y; and that when x is positive, that is in the first angle, x is an increasing function of y.

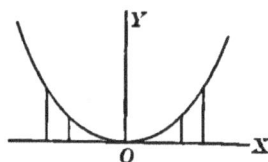

Fig. I.

Determine whether y is an increasing or a decreasing function of x in $y = \sin x$; $y = \tan x$; $y = \dfrac{1}{x}$; $y = a^x$; $y = \sqrt{a^2 - x^2}$.

4. Increments. *The amount of the increase or decrease of a variable in any interval of time is called its* **increment,** *or* **decrement.** It is usual, however, to employ the word increment to denote both an increase and a decrease, the increment receiving a negative sign where the variable is decreasing.

5. Uniform change. *A variable is said to change uniformly where its increment is numerically the same in all equal intervals of time.*

Since the increment is numerically the same for all equal intervals, the increment in *any* interval, assumed as a unit of time, may be taken as the measure of the change. This measure is called the **rate of change,** or simply the **rate,** of the variable, and is evidently constant. Hence *the rate of a uniformly changing variable is its increment in a unit of time.*

Representing by x the total change of the variable in the time t, and by r the change in the unit of time, $x = rt$ and

$$r = \frac{x}{t}; \tag{1}$$

or, *the rate of a uniformly changing variable is found by dividing the total change in any time t by t.*

6. Uniform motion. When the variable is the distance passed over by a moving point, estimated from any origin in the path, if this distance changes uniformly, the point is said to have uniform motion, and the increment of the distance in a unit of time is called the **velocity** of the point. Thus, if a point is said to have a velocity of 5 miles an hour, we mean that its distance from any point in its path increases or decreases 5 miles every hour. Hence *the velocity of a point having uniform motion is the rate of change of the distance it passes over.* Representing the distance passed over in the time

t by s, and by v the distance passed over in a unit of time, $v = \frac{s}{t}$, in which v is the rate of s.

7. Varied change. *When the law of change of a variable is such that in no two consecutive equal intervals of time its increments are equal, its change is said to be varied;* and the rate of such a variable at any instant is what its increment *would be* in a unit of time were the change at that instant to become uniform. Thus, if a point so moves that the increments of the distance passed over in consecutive equal intervals of time are unequal, its motion is said to be varied, and its velocity at any instant, that is, the rate of change of the distance, is the distance it *would* pass over in a unit of time were the motion to become uniform at that instant.

These definitions rest upon familiar conceptions. Suppose, for example, a cistern is being filled with water by a supply pipe in such a manner that the amount of water supplied is the same in all equal intervals of time, this amount being 5 gals. for one second. The quantity of water in the cistern (x) is a variable, and the amount of water *actually* supplied during any interval is its increment ; and because the change in x is uniform, we know not only the amount supplied in one second, but also in any other interval of time. For unequal intervals the corresponding increments are unequal, but the *rate* at which the cistern is being filled is the same throughout both intervals. The characteristic of uniform change is, therefore, a constant rate ; and we say the cistern is being filled at the rate of 5 gals. a second, or 300 gals. a minute, according as the second or the minute is the unit of time. If, now, the flow of water through the supply pipe ceases to remain uniform, the rate at which the cistern is being filled changes, the characteristic of varied change being a variable rate. In both cases the rate of the change of the quantity of water in the cistern is an instantaneous property of that quantity, but *in neither case can we measure it instantaneously.* When the flow is uniform, we observe what the *actual* change is for any definite interval ; when the flow varies, we ascertain what the change *would be* for any definite interval were the flow to become uniform at the instant considered. What these intervals are is immaterial ; but for the *comparison* of rates it is evidently necessary to adopt the same interval.

The following illustration is due to Clifford (Elements of Dynamic). Suppose a train to be moving from A to B on a straight track, its velocity being the same throughout the entire distance. Then its distance from A is a variable, and the distance passed over in any interval of time is the increment of the variable. If this increment is 20 miles for one hour, we know the increment for one minute will be $\frac{1}{3}$ of a mile, and that while these increments differ, the rate of change of x, the distance from A, is the same during the entire journey. If we now suppose a second train is moving in the same direction on a parallel track, and that it starts from A with a velocity less than 20 miles an hour, but gradually increasing to 40 miles an hour; and if we suppose further that its length is such that some part of it is always opposite to a traveller seated in the first train, then it will appear to him to be losing distance so long as its velocity is less than 20 miles an hour; but when its velocity exceeds 20 miles an hour, it will appear to be gaining. There must then be some instant between these two states of things at which the second train appears to the traveller to be neither gaining nor losing. At that instant the velocity of both trains is the same, i.e. 20 miles an hour, or the distance which the second train would pass over in one hour were its velocity at that instant to remain the same for one hour. In both cases, therefore, the velocity is determined in essentially the same manner; we suppose each train to maintain the velocity it has at any given instant for a unit of time and observe how far it goes. This increment is the rate at that instant.

8. Differentials. *What would be the increment of a variable in any interval of time were its rate to remain throughout the interval what it was at its beginning is called the* **differential** *of the variable.* It follows from Art. 4 that *the differential of a decreasing variable is negative.*

The symbol for the differential of any variable x is dx, read 'the differential of x.' The letter d must not be mistaken for a factor. Its meaning is 'the differential of,' as in $\sin x$ the abbreviation $\sin$ means 'the sine of.'

Since time (t) changes uniformly, any interval of time may be represented by dt.

9. Remark. The distinctions between the increment, differential, and rate, of a variable, should be carefully observed. Its increment is the amount of its *actual* increase, or decrease, in *any* interval of time; its differential is what the amount of its increase, or decrease, *would* be in any interval were its rate to remain throughout the interval what it was at its beginning. Hence the increment and differential of a variable are the same only when the variable is changing uniformly. Finally, its rate is what the amount of its increase, or decrease, would be *in a unit of time* were the change of the variable at any of its values to become uniform; a rate is thus a particular differential, namely, the differential for the unit of time.

10. **Corresponding increments,** or **differentials,** of variables are those which occur, or would occur, in the *same interval.*

Simultaneous rates of variables are their rates at the *same instant.*

The simultaneous rates of variables which are always equal are evidently equal.

11. Symbol of a rate. By Art. 7 the rate of any variable x at any instant, that is, at any of its values, is measured by the increment it would receive in a unit of time were its change at that instant to become *uniform;* hence if dx represents what this change would be in any interval dt, we have from Eq. (1), Art. 5,

$$r = \frac{dx}{dt}$$

whatever the interval dt; or the rate of any variable is the differential of the variable divided by the differential of t.

Cor. Since the differential is positive or negative as the variable is increasing or decreasing, *the rate of an increasing variable is positive, and of a decreasing variable is negative.*

Remark. It must be carefully noted that while dt is arbitrary, for the purpose of comparing the rates of different variables, or of the same variable at different instants, we must assume the same interval; hence dt *is constant.*

12. *Corresponding differentials of equals are equal.*

Let $y = f(x, z, v, \text{etc.})$. Since, when two quantities are always equal, their simultaneous rates are equal (Art. 10), if

dy and $d[f(x, z, v, \text{etc.})]$ be corresponding differentials of y and $f(x, z, v, \text{etc.})$, then

$$\frac{dy}{dt} = \frac{d[f(x, z, v, \text{etc.})]}{dt};$$

or, since dt is the same in both members, $dy = d[f(x, z, v, \text{etc.})]$. Hence, if an equation be true for all values of the variables involved, the corresponding differentials of the two members are equal.

$dy = d[f(x, z, v, \text{etc.})]$ is called the **first derived**, or **first differential, equation** of $y = f(x, z, v, \text{etc.})$.

The above is an immediate consequence of the definitions. For if a and β be any functions whatever, and $a = \beta$ for all values of the variables involved, the rates of a and β must be the same at any instant. Now these rates are what the changes in a and β would be in a unit of time were the common rate to become constant at any instant; and if the rate remained constant for *any* interval greater or less than the unit, the corresponding changes would still be equal; but these changes are the differentials.

13. *The immediate object of the Differential Calculus is the determination and comparison of the rates of variables.*

The following problem will serve as an illustration.

Suppose a wheel to revolve about a fixed axis through its centre, P being any point in the rim, and that we desire to compare the rate of P's motion in the arc AB with that of its motion vertically upward at any instant. This is equivalent to asking what are the rates of change of the arc AP and its sine

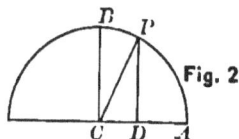

Fig. 2

PD. Hence if $AP = x$, $DP = y$, the fundamental relation is $y = \sin x$. Now if, as will be shown, $\dfrac{dy}{dt} = \cos x \dfrac{dx}{dt}$, the rate of y is seen to be $\cos x$ times the rate of x; that is, at any instant the sine is changing $\cos x$ times as fast as the arc. If P is moving in the arc at the rate of 10 ft. per sec., then at A, where $\cos x = 1$, it is also moving upward at the same rate.

At P, where $AP =$ arc of $60°$ and $\cos x = \cos 60° = \frac{1}{2}$, it is moving upward at the rate of 5 ft. per sec., or half as fast as it moves in the arc. At B, where $\cos x = \cos 90° = 0$, it is not moving upward at all.

14. Differentiation. In the above illustration the rate of y is the rate of $\sin x$; and, in general, the determination of the rates of variables involves the determination of the rates of the functions on which they depend or in which they enter. Since the rate of a variable is the differential of the variable divided by the differential of t, *the relation between the rates of variables will be known when the relation between their differentials is known.*

The process of determining the differential of a function is called **differentiation.**

We now proceed to determine rules for the differentiation of the several algebraic and transcendental functions.

CHAPTER II.

DIFFERENTIATION OF EXPLICIT FUNCTIONS.

THE ALGEBRAIC FUNCTIONS.

15. *The differential of a constant is zero.*

This is evident since a constant admits of no change, and therefore has no increment, whatever the interval. Properly speaking, such expressions as ' the differential of,' or ' rate of a constant' involve a contradiction of terms. But for uniformity of expression it is usual to say that both are zero.

16. *The differential of a polynomial is the algebraic sum of the differentials of its several terms.*

Let $y = x + z - v$. If the changes of x, z, and v, at any instant, that is, at any of their simultaneous values, become uniform, the change of y at that instant will also become uniform; and therefore, if dx, dz, dv, dy, be corresponding differentials of the variables and the function, $dy = dx + dz - dv$ (Art. 12). The above is evidently true of a polynomial of any number of terms.

Cor. $\frac{dy}{dt} = \frac{dx}{dt} + \frac{dz}{dt} - \frac{dv}{dt}$, or the rate of the sum of any number of variables is the sum of the rates of the variables.

Since the relation between the rates is always the same as that between the differentials, it will not be necessary to repeat this inference in the cases which follow.

17. *The differential of the product of a variable and a constant factor is the differential of the variable multiplied by the constant factor.*

Let $y = x + z + v +$ etc. From Art. 16, $dy = dx + dz + dv +$ etc. Hence if $x = z = v =$ etc., and m be the number of terms, $y = mx$ and $dy = dx + dx +$ etc. $= mdx$.

18. *The differential of the product of two variables is the sum of the products of each into the differential of the other.*

Let $y = xz$. Then y is the area of a rectangle whose sides are x and z. Let a, b, be any two simultaneous values of x and z; then at the instant when $x = a$ $= AB$ and $z = b = AD$, we have $y =$ area $ABCD$. Let BP represent what would be the change in x in the interval dt if at this instant its change were to become uniform, and DR the corresponding change in z were its change also to become uniform at the same instant. Then $BP = dx$, $DR = dz$. The change of y would then also become uniform, and for the interval dt would be $dy = BPQC + DRSC = bdx + adz$. But a and b are *any* simultaneous values of x and z. Hence, in general, at *any* instant, $dy = zdx + xdz$.

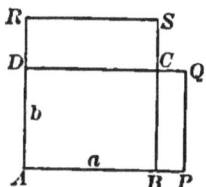

Fig. 3.

19. *The differential of the product of any number of variables is the sum of the products of the differential of each variable into all the others.*

Let $y = xzv$, and $xz = u$. Then $y = uv$. But $dy = vdu + udv$ (Art. 18), and $du = zdx + xdz$. Substituting in the former the value of du from the latter, and of $u = xz$, we have $dy = zvdx + xvdz + xzdv$.

In the same manner the theorem may be proved for the product of any number of variables.

20. *The differential of a fraction is the denominator into the differential of the numerator, minus the numerator into the differential of the denominator, divided by the square of the denominator.*

Let $y = \dfrac{x}{z}$. Then $x = yz$. But $dx = zdy + ydz$ (Art. 18).

Hence $dy = \dfrac{dx}{z} - y\dfrac{dz}{z} = \dfrac{dx}{z} - \dfrac{xdz}{z^2} = \dfrac{zdx - xdz}{z^2}$.

Cor. 1. If $x = a$, a constant, then $dx = 0$ (Art. 15), and $dy = -\dfrac{adz}{z^2}$; *or the differential of a fraction whose numerator is constant is minus the numerator into the differential of the denominator, divided by the square of the denominator.*

Cor. 2. If $z = a$, a constant, then $dz = 0$, and $dy = \dfrac{dx}{a}$, as it should be, since y is then $\dfrac{1}{a}x$ (Art. 17).

21. *The differential of a variable having a constant exponent is the product of the constant exponent, the variable with its exponent diminished by one, and the differential of the variable.*

I. *Let the exponent be positive and integral.*

Then $\qquad y = x^n = xxx \cdots$ to n factors.

Hence (Art. 19),
$$dy = x^{n-1}dx + x^{n-1}dx + \cdots \text{ to } n \text{ terms,}$$
or $\qquad dy = nx^{n-1}dx.$

II. *Let the exponent be a positive fraction.*

Then $y = x^{\frac{m}{n}}$, whence $y^n = x^m$. The differentials of the two numbers of this equation being equal (Art. 12), we have, by I.,
$$ny^{n-1}dy = mx^{m-1}dx,$$
whence $\qquad dy = \dfrac{m}{n}\dfrac{x^{m-1}}{y^{n-1}}dx = \dfrac{m}{n}x^{\frac{m}{n}-1}dx.$

III. *Let the exponent be negative.*

Then $y = x^{-n} = \dfrac{1}{x^n}$, n being fractional or integral. Clearing of fractions, $yx^n = 1$; whence, differentiating the product and remembering that the differential of a constant is zero,

$$x^n \, dy + ny x^{n-1} \, dx = 0,$$

or $$dy = -\frac{ny x^{n-1} \, dx}{x^n} = - nx^{-n-1} \, dx.$$

Cor. 1. If $n = \frac{1}{2}$, $y = \sqrt{x}$, and $dy = \dfrac{dx}{2\sqrt{x}}$, or *the differential of the square root of a variable is the differential of the variable divided by twice the square root of the variable.*

Special rules might be framed for $n = \frac{1}{3}$, $n = \frac{1}{4}$, etc., but in such cases the general rule is preferable.

Remark. The above method of proof depends upon the resolution of the power into equal factors, and is therefore inapplicable to the case of a variable having an imaginary, or an incommensurable, exponent. The rule, however, as will subsequently appear, holds good for these cases also.

Examples. Differentiate:

1. $y = x^2 + 3x - 4x^3$.
$dy = d(x^2 + 3x - 4x^3) = d(x^2) + d(3x) - d(4x^3)$ [Art. 16
$= 2x \, dx + 3 \, dx - 12 x^2 dx = (2x + 3 - 12 x^2) \, dx$.
[Arts. 21, 15

2. $y = a + mx^m - 7 nx^2$. $\quad dy = (m^2 x^{m-1} - 14 nx) \, dx$.

3. $y^2 = 2px$.

Although an implicit function, it may be differentiated directly without first reducing to an explicit form. Thus,

$d(y^2) = d(2px)$, whence $2y \, dy = 2p \, dx$, or $dy = \dfrac{p}{y} \, dx$.

4. $a^2 y^2 \pm b^2 x^2 = \pm a^2 b^2$. $\quad dy = \mp \dfrac{b^2 x}{a^2 y} \, dx$.

5. $y = (1 + x^2)(1 - 2x^3)$.
$dy = (1 - 2x^3) d(1 + x^2) + (1 + x^2) d(1 - 2x^3)$
$= (1 - 2x^3) 2x \, dx - (1 + x^2) 6 x^2 dx$
$= 2(x - 3x^2 - 5x^4) dx$; or we may first expand and then differentiate.

6. $y = (a + bx^3)^{\frac{1}{2}}$.

$dy = d[(a + bx^3)^{\frac{1}{2}}] = \frac{1}{2}(a + bx^3)^{-\frac{1}{2}} d(a + bx^3)$ [Art. 21

$= \dfrac{3bx^2 dx}{2(a + bx^3)^{\frac{1}{2}}}$; or, more expeditiously by the special

rule for the square root of a variable (Art. 20).

7. $y = (ax)^{\frac{1}{2}} + bx^{\frac{2}{3}}$. $dy = \left(\dfrac{1}{2} \sqrt{\dfrac{a}{x}} + \dfrac{2}{3} \dfrac{b}{x^{\frac{1}{3}}} \right) dx$.

8. $y = (1 + x)\sqrt{1 - x}$. $dy = \dfrac{1 - 3x}{2\sqrt{1 - x}} dx$.

9. $y = \sqrt{1 + \sqrt{x}}$. $dy = \dfrac{dx}{4\sqrt{x}\sqrt{1 + \sqrt{x}}}$.

10. $y = \dfrac{x^n v}{z^2}$.

Put the expression in the form $x^n z^{-2} v$.   Then

$$dy = \dfrac{nvx^{n-1}dx}{z^2} - \dfrac{2x^n v\, dz}{z^3} + \dfrac{x^n dv}{z^2}.$$

11. $y = \sqrt{xz^2 v^{\frac{1}{2}}}$.

$dy = \dfrac{d(xz^2 v^{\frac{1}{2}})}{2\sqrt{xz^2 v^{\frac{1}{2}}}} = \dfrac{z^2 v^{\frac{1}{2}}dx + 2xv^{\frac{1}{2}}zdz + \frac{1}{2}xz^2 v^{-\frac{1}{2}}dv}{2x^{\frac{1}{2}} zv^{\frac{1}{4}}}$

$= \dfrac{zv^{\frac{1}{4}}dx}{2\sqrt{x}} + v^{\frac{1}{4}}\sqrt{x}\, dz + \dfrac{z\sqrt{x}\, dv}{4v^{\frac{3}{4}}}$.

12. $y = \dfrac{1}{x}$. $dy = -\dfrac{dx}{x^2}$.

13. $y = \dfrac{a}{\sqrt{x}}$. $dy = -\dfrac{adx}{2x^{\frac{3}{2}}}$.

14. $y = \dfrac{\sqrt{x}}{2}$. $dy = \dfrac{dx}{4\sqrt{x}}$.

15. $y = \dfrac{1 + x}{1 - x}$. $dy = \dfrac{2\, dx}{(1 - x)^2}$.

$dy = \dfrac{(1 - x)d(1 + x) - (1 + x)d(1 - x)}{(1 - x)^2}$.

16. $y = \dfrac{x}{a - x}$. $dy = \dfrac{adx}{(a - x)^2}$.

17. $y = \dfrac{x^n}{(1+x)^n}.$ $\qquad dy = \dfrac{nx^{n-1}}{(1+x)^{n+1}}dx.$

18. $y = \left(\dfrac{x}{1-x}\right)^2.$ $\qquad dy = \dfrac{2\,x dx}{(1-x)^3}.$

19. $y = \dfrac{2\,abx^4}{(1+ax^2)^{\frac{5}{2}}}.$ $\qquad dy = \dfrac{2\,abx^3(4-ax^2)\,dx}{(1+ax^2)^{\frac{7}{2}}}.$

20. $y = \dfrac{2+mx+x^3}{x}.$ $\qquad dy = 2\left(x - \dfrac{1}{x^2}\right)dx.$

Put the expression in the form $y = \dfrac{2}{x} + m + x^2.$

21. $y = \dfrac{2}{\sqrt{1-x^2}}.$ $\qquad dy = \dfrac{2\,x dx}{\sqrt{(1-x^2)^3}}.$

Put the expression in the form $y = 2(1-x^2)^{-\frac{1}{2}}$ and differentiate as a power rather than as a fraction.

22. $y = \dfrac{\sqrt{1+x^2}}{x}.$ $\qquad dy = -\dfrac{dx}{x^2\sqrt{1+x^2}}.$

23. $y = \dfrac{x^3}{(1-x^2)^{\frac{3}{2}}}.$ $\qquad dy = \dfrac{3\,x^2 dx}{(1-x^2)^{\frac{5}{2}}}.$

24. $y = \sqrt{\dfrac{1-x}{1+x}}.$ $\qquad dy = -\dfrac{dx}{(1+x)\sqrt{1-x^2}}.$

25. $y = \sqrt{1 + \dfrac{a}{x} + \dfrac{b}{x^2}}.$ $\qquad dy = -\dfrac{1}{2\,x^2}\dfrac{ax+2b}{\sqrt{x^2+ax+b}}\,dx.$

26. $y = \dfrac{\sqrt{a+x}}{\sqrt{a}+\sqrt{x}}.$ $\qquad dy = \dfrac{\sqrt{a}(\sqrt{x}-\sqrt{a})\,dx}{2\sqrt{x}\sqrt{a+x}\,(\sqrt{a}+\sqrt{x})^2}.$

27. $y = \dfrac{x^2}{\sqrt{1+x^2}-x}.$ $\qquad dy = \left\{\dfrac{2x+3\,x^3}{\sqrt{1+x^2}} + 3x^2\right\}dx.$

Rationalize the denominator before differentiating.

28. $y = \dfrac{\sqrt{1-x^2}-x}{\sqrt{1-x^2}+x}.$ $\qquad dy = 2\left\{\dfrac{2\,x\sqrt{1-x^2}-1}{(1-2x^2)^2\sqrt{1-x^2}}\right\}dx.$

29. $y = \dfrac{1-x}{\sqrt{x}}.$ $\qquad dy = -\dfrac{1+x}{2\,x^{\frac{3}{2}}}\,dx.$

30. $y = \dfrac{1}{\sqrt{1+x}} + \dfrac{1}{\sqrt{1-x}}.$ $dy = \dfrac{1}{2}\left(\dfrac{1}{(1-x)^{\frac{3}{2}}} - \dfrac{1}{(1+x)^{\frac{3}{2}}}\right)dx.$

31. $y = \dfrac{c}{x - \sqrt{x^2 - c}}.$ $dy = \left(1 + \dfrac{x}{\sqrt{x^2 - c}}\right)dx.$

32. $y = \dfrac{a}{(b+x)^m (c+x)^n}.$ $dy = -\,a\,\dfrac{m(c+x) + n(b+x)}{(b+x)^{m+1}(c+x)^{n+1}}\,dx.$

33. $y = \dfrac{1}{x + \sqrt{1 - x^2}}.$ $dy = \dfrac{x - \sqrt{1-x^2}}{\sqrt{1-x^2}\,(1 + 2x\sqrt{1-x^2})}\,dx.$

34. $y = \sqrt{\dfrac{1-x^2}{(1+x^2)^3}}.$ $dy = -\,\dfrac{2x\,(2-x^2)}{(1-x^2)^{\frac{1}{2}}(1+x^2)^{\frac{5}{2}}}\,dx.$

35. $y = (2a^{\frac{1}{2}} + x^{\frac{1}{2}})(a^{\frac{1}{2}} + x^{\frac{1}{2}})^{\frac{1}{2}}.$ $dy = \dfrac{4a^{\frac{1}{2}} + 3x^{\frac{1}{2}}}{4x^{\frac{1}{2}}(a^{\frac{1}{2}} + x^{\frac{1}{2}})^{\frac{1}{2}}}\,dx.$

36. $y = (x - \sqrt{1-x^2})^n.$ $dy = n(x - \sqrt{1-x^2})^{n-1}\dfrac{\sqrt{1-x^2}+x}{\sqrt{1-x^2}}\,dx.$

37. $y = \dfrac{\sqrt{1+x^2} + \sqrt{1-x^2}}{\sqrt{1+x^2} - \sqrt{1-x^2}}.$ $dy = -\dfrac{2}{x^3}\left(1 + \dfrac{1}{\sqrt{1-x^4}}\right)dx.$

22. Analytic signification of the ratio $\dfrac{dy}{dx}$.

Let $y = f(x)$. The only variable which enters the function being x, the function y will change only as its variable x changes, and the rate of change of y will depend upon the rate of change of x. Let k be the ratio of these rates at any instant. Then

$$\frac{dy}{dt} = k\frac{dx}{dt}, \text{ whence } \frac{dy}{dx} = k.$$

Hence *the ratio of the differential of the function to the differential of the variable is the ratio of the rate of change of the function to that of the variable.* It is evidently independent of the interval dt.

As derived by differentiation from the function $y = f(x)$, this ratio is called its **first derived function**, or simply its **first derivative**; also, being the factor k by which the rate of the variable is multiplied to obtain the rate of the function, it is called the **first differential coefficient**.

Cor. If $\dfrac{dy}{dx}$ is positive, y is an increasing function of x; and if negative, y is a decreasing function of x (Art. 11, Cor.).

23. Applications. 1. Compare the rates of change of the ordinate and abscissa of the parabola whose parameter is 4. Which is changing most rapidly at the point $x = 9$? Where are they changing at the same rate? If at the point $x = 16$ the abscissa is increasing at the rate of 24 ft. a second, at what rate is the ordinate then changing?

$y^2 = 4x$, $\therefore \dfrac{dy}{dx} = \dfrac{2}{y}$, which may be written $\dfrac{dy}{dt} = \dfrac{2}{y}\dfrac{dx}{dt}$; or, in general, the ordinate changes $\dfrac{2}{y}$ times as fast as the abscissa. For $x = 9$, $y = \pm 6$, and $\dfrac{2}{y}$ becomes $\pm \dfrac{1}{3}$, showing that the ordinate is increasing, or decreasing, $\dfrac{1}{3}$ as fast as the abscissa at the points $(9, 6)$, $(9, -6)$. When the ordinate and abscissa are changing at the same rate, we must have $\dfrac{dy}{dx} = \dfrac{2}{y} = 1$, $\therefore y = 2$, which, in the equation of the curve, gives $x = 1$. This is as it should be, for $(1, 2)$ is the extremity of the parameter, at which point the generating point is moving in the direction of the focal tangent, whose inclination to the axis of X is known to be 45°. For $x = 16$, $y = \pm 8$, and $\dfrac{2}{y}$ becomes $\pm \dfrac{1}{4}$, or $\dfrac{dy}{dt} = \pm \dfrac{1}{4}\dfrac{dx}{dt}$; hence if at $x = 16$ the abscissa is increasing at the rate of 24 ft. a second, the ordinate in the first angle is increasing, and in the fourth angle decreasing, at the rate of $\frac{1}{4} \times 24 = 6$ ft. a second.

2. Compare the rates of change of x and y in the ellipse $a^2y^2 + b^2x^2 = a^2b^2$. Is y an increasing or decreasing function of

x? Compare the rates at the extremities of the axes. If the axes are 6 and 4, at what point is y changing $2\sqrt{\dfrac{5}{3}}$ as fast as x?

$\dfrac{dy}{dx} = -\dfrac{b^2x}{a^2y}$, and is negative when x and y have like signs; hence the function is decreasing in the first and third, and increasing in the second and fourth, angles. At the extremities of the transverse axis, $y = 0$, and $\dfrac{dy}{dx} = \infty$, or y is changing infinitely faster than x. To determine the point where y changes $2\sqrt{\dfrac{5}{3}}$ as fast as x, we have $-\dfrac{b^2x}{a^2y} = -\dfrac{4x}{9y} = 2\sqrt{\dfrac{5}{3}}$, which, with the equation of the curve $9y^2 + 4x^2 = 36$, gives four points whose coordinates are numerically $\dfrac{3}{4}\sqrt{15}$ and $\dfrac{1}{2}$.

3. The altitude of a right triangle increases at the rate of 10 in. a second. At what rate is the area increasing?

Let $b = $ base, $x = $ altitude, $y = $ area. Then $y = \dfrac{bx}{2}$, $\therefore \dfrac{dy}{dx} = \dfrac{b}{2}$, which is a constant; therefore the area increases uniformly at the rate of $\dfrac{b}{2} \times 10 = 5b$ sq. in. a second.

4. A spherical balloon is being filled with gas at the rate of m cub. ft. a second. At what rate is the diameter increasing when its length is 6 ft.?

Let $y = $ diameter, $x = $ volume.

Then $x = \dfrac{\pi}{6}y^3$, or $y = \left(\dfrac{6}{\pi}\right)^{\frac{1}{3}} x^{\frac{1}{3}}$, $\therefore \dfrac{dy}{dx} = \sqrt[3]{\dfrac{2}{9\pi x^2}}$.

When $y = 6$, $x = 36\pi$, and $\dfrac{dy}{dx} = \dfrac{1}{18\pi}$.

Hence when the diameter is 6, it is increasing at the rate of $\dfrac{m}{18\pi}$ ft. a second.

5. A rectangle whose sides are parallel to the axes is inscribed in the ellipse $a^2y^2 + b^2x^2 = a^2b^2$. Compare the rates of change of its area and side.

Let x and y be the half sides and z the area. Then $z = 4xy$. To eliminate y, and so obtain z a function of a single variable, we have from the equation of the ellipse,

$$y = \frac{b}{a}\sqrt{a^2 - x^2}, \ \therefore \ z = \frac{4b}{a}\sqrt{a^2 x^2 - x^4}, \ \frac{dz}{dx} = \frac{4b}{a}\frac{a^2 - 2x^2}{\sqrt{a^2 - x^2}}.$$

In a similar manner we may find the ratio $\dfrac{dz}{dy}$.

6. The radius and altitude of a right cone vary, the slant height remaining constantly 25 ft. Compare the rates of change of the volume and altitude. When the altitude is 4 ft. and changing 3 ft. a second, how fast is the volume changing?

Let $s =$ slant height, $x =$ altitude, $z =$ radius of base, and $y =$ volume. Then $y = \dfrac{\pi z^2 x}{3}$. To eliminate z we have the condition $z^2 + x^2 = s^2$, $\therefore$ $z^2 = s^2 - x^2$ and $y = \dfrac{\pi}{3}(s^2 x - x^3)$. Whence $\dfrac{dy}{dx} = \dfrac{\pi}{3}(s^2 - 3x^2) = \dfrac{\pi}{3}(625 - 3x^2)$; that is, the volume is increasing $\dfrac{\pi}{3}(625 - 3x^2)$ times as fast as the altitude is in linear feet. When $x = 4$, this becomes $\dfrac{577}{3}\pi$, and at that instant the volume is changing at the rate of 577π cub. ft. a second.

The student will observe that he may eliminate before or after differentiation. Thus, differentiating first,

$$dy = \frac{\pi}{3}(2zxdz + z^2 dx), \ \therefore \ \frac{dy}{dx} = \frac{\pi}{3}\left(2zx\frac{dz}{dx} + z^2\right).$$

But from $z^2 + x^2 = s^2$, $\dfrac{dz}{dx} = -\dfrac{x}{z}$; substituting this value with those of z and z^2, $\dfrac{dy}{dx} = \dfrac{\pi}{3}(s^2 - 3x^2)$, as before.

7. A point P moves from A at a uniform rate in the direction of AP, at right angles to AB. A light C, whose intensity at a unit's distance is 125, is vertically over B. If $AB = 10$, $BC = 5$, compare the rates of the motion and illumination of

P (understanding that the intensity of a light at any point is its intensity at a unit's distance divided by the square of the distance), when $AP = 10$.

Let $AP = x$. Then the illumination at

$$P = y = \frac{125}{CP^2} = \frac{125}{10^2 + 5^2 + x^2},$$

$$\therefore \frac{dy}{dx} = -\frac{250\,x}{(125 + x^2)^2},$$

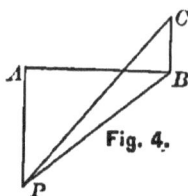

Fig. 4.

which, when $x = 10$, becomes $-\frac{4}{81}$, or the rate of change of the illumination of P is $\frac{4}{81}$ times the rate of change of AP, and is decreasing.

8. If, in Fig. 4, BC is a lamp-post 10 ft. high, and a man whose height is 6 ft. walks from B in the direction BA at the rate of 3 miles an hour, show that the extremity of his shadow moves at the rate of $7\frac{1}{2}$ miles an hour.

9. In Ex. 8 show that the length of the man's shadow is increasing at the rate of $4\frac{1}{2}$ miles an hour.

10. In Fig. 4 show that if the man walks from A towards B, he is approaching B $\frac{y}{x}$ times as fast as he is approaching C, where $BA = x$, $CA = y$.

11. A ship is sailing northeast at the rate of 10 miles an hour. At what rate is it making north latitude?

Ans. $5\sqrt{2}$ miles an hour.

12. The area of a circular plate of metal is expanded by heat. Find the rate of change of the area when the radius is 5 in. and increasing .01 in. a sec. Ans. $\frac{1}{10}\pi$ sq. in. a sec.

13. If the thickness of the plate of Ex. 12 increases one-half as fast as the radius, find the rate of increase of the volume when the radius is 5 in. and the thickness .5 in.

Let $v = $ volume, $x = $ radius, $y = $ thickness. Then $v = \pi x^2 y$; whence

$$\frac{dv}{dt} = 2\pi xy\frac{dx}{dt} + \pi x^2 \frac{dy}{dt} = (2\pi xy + \tfrac{1}{2}\pi x^2)\frac{dx}{dt} = \frac{7}{40}\pi \text{ cub. in. a sec.}$$

14. Two ships, on courses whose included angle is 60°, are sailing away from the intersection of the courses with velocities of 6 and 4 miles an hour. Find the rate at which they are separating when 10 and 15 miles respectively from the intersection.

Let $z = $ distance between the ships, x and y their distances from the intersection, 6 and 4 being the rates of x and y respectively.

Then $z^2 = x^2 + y^2 - 2xy\cos 60° = x^2 + y^2 - xy$,

$$\therefore \frac{dz}{dt} = \frac{1}{2z}\left[(2x - y)\frac{dx}{dt} + (2y - x)\frac{dy}{dt}\right] = \frac{55}{\sqrt{175}}.$$

15. Find the rate of separation in Ex. 14 under the supposition that the ships start together from the intersection of the courses, with the velocities 6 and 4.

$$z^2 = (6t)^2 + (4t)^2 - 24t^2, \quad \therefore \frac{dz}{dt} = \sqrt{28}.$$

16. C is any point without a circle whose centre is O, and OC cuts the circle at A. Find the relative rates of departure from C and OC of a point P moving from A in the arc of the circle.

Let P be the position of the point at any instant, $y = PM$, the perpendicular on OC, $PC = x$, $OC = a$, $OA = R$. Then

$$PC = \sqrt{CM^2 + PM^2},$$

or $$x = \sqrt{(a - \sqrt{R^2 - y^2})^2 + y^2},$$

$$\therefore dx = \frac{aydy}{x\sqrt{R^2 - y^2}}.$$

24. Geometric signification of $\dfrac{dy}{dx}$.

Since to every equation $y = f(x)$ there corresponds some plane locus, the ratio $\dfrac{dy}{dx}$ is evidently capable of geometric interpretation.

Let $M'N'$ be the locus of $y = f(x)$, and P the position of the generating point at any instant. Then dx, dy, being corresponding differentials of x and y, are what would be the changes in $x = OD$ and $y = DP$ during any interval if at its beginning their rates of change should become constant. But this will evidently be the case if at P the motion of the generating point should become uniform along the tangent at P. Hence PQ, QR, being

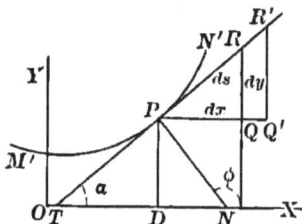

Fig. 5.

what would be the corresponding increments of x and y in any interval dt if the change of each became uniform at the instant considered, are corresponding differentials of x and y, and

$$\frac{QR}{PQ} = \frac{dy}{dx} = \tan XTP = \tan a. \tag{1}$$

The tangent of the angle made by any straight line with the axis of X is called the **slope** of the line. As the tangent at any point of a curve has the direction of the curve at that point, the slope of a curve is that of its tangent; hence *the value of* $\dfrac{dy}{dx}$ *at any instant, that is, for any simultaneous values of x and y, measures the slope of the curve at the corresponding point.*

In the figure, y is an increasing function of x, a is an acute angle, and $\tan a$, or $\dfrac{dy}{dx}$, is positive. In the vicinity of M', however, y is a decreasing function of x, a is an obtuse angle, and $\tan a$, or $\dfrac{dy}{dx}$, is negative, as already seen in Art. 22.

It is evident that the slope will in general vary from point to point, and the first derivative will therefore be in general a function of x; but that for any particular value of x it has a definite value independent of dt, that is, independent of dx, since from the similar triangles PQR, $PQ'R'$, $\dfrac{dy}{dx}$ remains constant, whatever the interval.

25. Relations between the velocities in the path and along the axes.

Let $s =$ distance passed over by the generating point, estimated from any point in its path, that is, the length of the path. Since, when the changes in x and y (Fig. 5) become uniform, the generating point moves in the direction of the tangent PR, $PR = ds$, $PQ = dx$, $QR = dy$, are corresponding differentials of s, x, and y; and from the right triangle PQR,

$$ds^2 = dx^2 + dy^2. \tag{1}$$

Hence if $y = f(x)$ be the path of a moving point, $\dfrac{ds}{dt}$ is the rate of change of the distance, or *the velocity of the point in its path;* and, for like reasons, $\dfrac{dx}{dt}$, $\dfrac{dy}{dt}$, are its velocities in the directions of the axes.

By differentiating $y = f(x)$ we can compare the horizontal and vertical velocities, and substituting either $\dfrac{dx}{dt}$ or $\dfrac{dy}{dt}$ from the differential equation of the path in

$$\frac{ds}{dt} = \sqrt{\left(\frac{dx}{dt}\right)^2 + \left(\frac{dy}{dt}\right)^2},$$

we can compare the velocity in the path with either the horizontal or vertical velocity.

Since $\dfrac{dx}{dt}$ and $\dfrac{dy}{dt}$ are distances, namely, the distances which the point would pass over in a unit of time in the directions of the axes if its velocity in each direction became uniform, they are positive or negative according as each is in the positive or negative direction of the corresponding axis.

26. The following relations will be found of use hereafter. Let PN (Fig. 5) be the normal at P. Then

$$\cos a = \quad \sin \phi = \frac{dx}{ds}, \tag{1}$$

$$\sin a = -\cos \phi = \frac{dy}{ds}. \tag{2}$$

27. Applications. 1. To find the general equation of a tangent to any plane curve.

Let $y = f(x)$ be the equation of the curve and (x', y') the point of tangency. The equation of a straight line through (x', y') is $y - y' = m(x - x')$. If we form $\frac{dy}{dx}$ from the equation of the curve, and substitute in it the coordinates of the given point of tangency, we have the slope of the curve at this point (Art. 24). But the slope of a curve at any point is that of its tangent at that point; hence, representing by $\frac{dy'}{dx'}$ what $\frac{dy}{dx}$ becomes for the point (x', y'), and substituting $\frac{dy'}{dx'}$ for m,

$$y - y' = \frac{dy'}{dx'}(x - x').$$

2. Deduce the equation of the tangent to the ellipse $a^2y^2 + b^2x^2 = a^2b^2$.

From the equation of the ellipse, $\frac{dy}{dx} = -\frac{b^2x}{a^2y}$, the general expression for the slope. For the particular point (x', y') this becomes $-\frac{b^2x'}{a^2y'}$, and the equation of the tangent is, therefore,

$y - y' = -\frac{b^2x'}{a^2y'}(x - x')$. Clearing of fractions and substituting for $a^2y'^2 + b^2x'^2$ its value a^2b^2, the equation assumes the simpler form $a^2yy' + b^2xx' = a^2b^2$.

Show that the equation of the tangent to:

3. The hyperbola $a^2y^2 - b^2x^2 = -a^2b^2$ is $a^2yy' - b^2xx' = -a^2b^2$.

4. The parabola $y^2 = 2px$ is $yy' = p(x + x')$.

5. The circle $y^2 + x^2 = R^2$ is $yy' + xx' = R^2$.

6. The circle $y^2 = 2Rx - x^2$ is $y - y' = \dfrac{R - x'}{y'}(x - x')$.

7. The hyperbola $xy = m$, referred to its asymptotes, is $y = -\dfrac{y'}{x'}x + 2y'$.

8. The cissoid $y^2 = \dfrac{x^3}{2a - x}$ is $y - y' = \pm \dfrac{x'^{\frac{1}{2}}(3a - x')}{(2a - x')^{\frac{3}{2}}}(x - x')$.

9. The curve $a^3y^3 + b^3x^3 = a^3b^3$ is $y - y' = -\dfrac{b^3x'^2}{a^3y'^2}(x - x')$.

10. Find the slope of $y^2 = 2px$ at the vertex; at the extremities of the parameter. Is the generating point ever moving in a direction parallel to X?

From $y^2 = 2px$, $\dfrac{dy}{dx} = \dfrac{p}{y}$, which is ∞ for $y = 0$. Hence the tangent is perpendicular to X at the vertex. For $y = \pm p$, $\dfrac{dy}{dx} = \pm 1$, the slope of the focal tangents, which therefore make angles of 45° and 135° with X. Since $\dfrac{dy}{dx}$ is zero only when $y = \infty$, $\therefore x = \infty$, there is no point at which the tangent is parallel to X.

11. Find the slope of $y = x^4 - 2x^2 + 3$ at $x = 1$; $x = 3$; $x = -2$. *Ans.* 0; 96; −24.

12. At what point of $y^2 = ax^3$ is the slope 0? 1?

Ans. $(0, 0)$; $\left(\dfrac{4}{9a}, \dfrac{8}{27a}\right)$.

13. Find the equation of the tangent to the parabola $y^2 = 2px$ inclined 30° to X.

Since the angle is 30°, its slope is $\dfrac{1}{\sqrt{3}}$, and we must have $\dfrac{p}{y'} = \dfrac{1}{\sqrt{3}}$, or $y' = p\sqrt{3}$. Hence from the equation of the curve, $x' = \dfrac{3}{2}p$. Substituting these values in $yy' = p(x + x')$, we obtain $y = \dfrac{1}{\sqrt{3}}x + \dfrac{\sqrt{3}}{2}p$.

14. At what angle does $y^2 = 12x$ intersect $y^2 + x^2 + 6x - 63 = 0$?

The points of intersection are $(3, \pm 6)$; the slopes are $\dfrac{6}{y}$, $-\dfrac{3+x}{y}$, which become, for the point $(3, 6)$, 1 and -1. These being negative reciprocals of each other, the curves intersect at $(3, 6)$ at an angle 90°.

15. Show that the cissoid cuts its circle at an angle whose tangent is 2.

16. Show that the length of the tangent to the hypocycloid $x^{\frac{2}{3}} + y^{\frac{2}{3}} = a^{\frac{2}{3}}$ intercepted between the coordinate axes is constant.

The equation of the tangent is $\dfrac{x}{x'^{\frac{1}{3}}} + \dfrac{y}{y'^{\frac{1}{3}}} = a^{\frac{2}{3}}$.

17. To find the general equation of the normal to any plane curve.

The normal passes through the point of contact and is perpendicular to the tangent. The condition of perpendicularity is $m' = -\dfrac{1}{m}$. Hence the equation of the normal is $y - y' = -\dfrac{dx'}{dy'}(x - x')$.

18. Find the equations of the normals to the conic sections:

Ellipse, $y - y' = \dfrac{a^2 y'}{b^2 x'}(x - x')$; Parabola, $y - y' = -\dfrac{y'}{p}(x - x')$;

Circle, $y = \dfrac{y'}{x'}x$; Hyperbola, $y - y' = -\dfrac{a^2 y'}{b^2 x'}(x - x')$.

19. Find the equations of the tangent and normal to $y^2 = 9x^3$ at the point (1, 3). *Ans.* $y = \frac{9}{2}x - \frac{3}{2}$; $y = -\frac{2}{9}x + \frac{29}{9}$.

20. To find the lengths of the subtangent and tangent to any plane curve.

In Fig. 5,

$$TD = \frac{PD}{\tan DTP} = \frac{y'}{\frac{dy'}{dx'}} = y'\frac{dx'}{dy'}.$$

$$TP = \sqrt{TD^2 + DP^2} = \sqrt{y'^2 + \left(y'\frac{dx'}{dy'}\right)^2} = y'\sqrt{1 + \left(\frac{dx'}{dy'}\right)^2}.$$

21. To find the lengths of the subnormal and normal to any plane curve.

In Fig. 5, $DN = PD\tan DPN = PD\tan DTP = y'\frac{dy'}{dx'}.$

$$NP = y'\sqrt{1 + \left(\frac{dy'}{dx'}\right)^2}.$$

22. Find the subtangents and subnormals of the conic sections.

	SUBTANGENT.	SUBNORMAL.
Ellipse,	$\dfrac{x'^2 - a^2}{x'}.$	$-\dfrac{b^2 x'}{a^2}.$
Hyperbola,	$\dfrac{x'^2 - a^2}{x'}.$	$\dfrac{b^2 x'}{a^2}.$
Circle,	$-\dfrac{y'^2}{x'}.$	$-x'.$
Parabola,	$2x'.$	$p.$

The signs may be neglected if lengths only are required. The sign will, however, indicate the direction if the subtangent and subnormal be reckoned respectively from T and D, Fig. 5.

23. Prove that the subtangent of the hyperbola $xy = m$ is the abscissa of the point of contact, and that the subnormal varies as the cube of the ordinate.

$$y'\frac{dx'}{dy'} = -x'; \quad y'\frac{dy'}{dx'} = -\frac{y'^3}{m}.$$

24. Prove that the subtangent of the semi-cubical parabola $y^2 = ax^3$ is two thirds the abscissa of the point of contact, and that the subnormal varies directly as the square of the abscissa.

25. A point moves with a constant velocity m in the arc of the parabola $y^2 = 8x$. Find the velocities in the directions of the axes when $x = 8$.

From $y^2 = 8x$, we have $\dfrac{dy}{dt} = \dfrac{4}{y}\dfrac{dx}{dt}$, and by condition $\dfrac{ds}{dt} = m$. Substituting these values in

$$\frac{ds}{dt} = \sqrt{\left(\frac{dx}{dt}\right)^2 + \left(\frac{dy}{dt}\right)^2},$$

we obtain $\dfrac{dx}{dt} = \dfrac{my}{\sqrt{y^2 + 16}}, \qquad \therefore \dfrac{dy}{dt} = \dfrac{4}{y}\dfrac{dx}{dt} = \dfrac{4m}{\sqrt{y^2 + 16}}.$

For $x = 8$, $\therefore y = 8$, these become $\dfrac{2m}{\sqrt{5}}$ and $\dfrac{m}{\sqrt{5}}$; hence at the point $(8, 8)$ the horizontal and vertical velocities are as 2 to 1, and are $\dfrac{2}{\sqrt{5}}$ and $\dfrac{1}{\sqrt{5}}$ times that in the path.

26. The orbit of a comet is a parabola, the sun occupying the focus. Compare the velocity of the comet with its rate of approach to the sun.

The distance of any point of the parabola from the focus is $r = x + \dfrac{p}{2}$, $\therefore \dfrac{dr}{dt} = \dfrac{dx}{dt}$, or its rate of approach to the sun is the same as its horizontal velocity. But, as shown in Ex. 25, $\dfrac{dx}{dt} = \dfrac{y}{\sqrt{y^2 + p^2}}\dfrac{ds}{dt}$. Hence, in general, its rate of approach to the sun is $\dfrac{y}{\sqrt{y^2 + p^2}}$ times its velocity. At the vertex, $y = 0$ and $\dfrac{dx}{dt} = \dfrac{dr}{dt} = 0$, or, at the vertex, it is not approaching the sun at all. When $y = p$, $\dfrac{dr}{dt} = \dfrac{1}{\sqrt{2}}\dfrac{ds}{dt}$. When at a distance from the sun equal to the parameter of the orbit,

$$r = 2p = x + \frac{p}{2}, \quad \therefore x = \frac{3}{2}p \text{ and } y = \sqrt{3}p, \text{ and } \frac{dr}{dt} = \frac{1}{2}\sqrt{3}\frac{ds}{dt}.$$

27. A point moves in the arc of the circle $x^2 + y^2 = 25$, and has a velocity 10 in passing through the point $(3, 4)$. Show that its velocities in the directions of the axes are 8 and 6, numerically.

THE TRANSCENDENTAL FUNCTIONS.

The Logarithmic and Exponential Functions.

28. The logarithmic function.

Let $\qquad x = ny,$ $\qquad\qquad(1)$

n being any arbitrary constant. Then

$$\log_a x = \log_a n + \log_a y, \qquad\qquad(2)$$

in which a is the base of the logarithmic system.

Differentiating (1) and (2),

$$dx = ndy,$$

$$d(\log_a x) = d(\log_a y);$$

and, by division, $\qquad \dfrac{d(\log_a x)}{dx} = \dfrac{d(\log_a y)}{ndy}. \qquad\qquad(3)$

Eliminating n from (3) by substituting its value from (1),

$$\frac{d(\log_a x)}{\dfrac{dx}{x}} = \frac{d(\log_a y)}{\dfrac{dy}{y}}, \qquad\qquad(4)$$

or the ratio of $d(\log_a x)$ to $\dfrac{dx}{x}$ is the same as that of $d(\log_a y)$ to $\dfrac{dy}{y}$. Since n is arbitrary, the ratios in (4) are constant. Let m be this constant. Then

$$d(\log_a x) = m\frac{dx}{x}.$$

Now the only quantities involved in any logarithmic system are the number, its logarithm, and the base. Since of these the two former are variable, while m is constant, m must depend upon the base. The value of m corresponding to any base is called the **modulus** of that system. Hence

The differential of a logarithm of a variable is the modulus of the system into the differential of the variable divided by the variable.

The relation between the modulus and the base of any system will be established later ; but as the only system employed in analytic investigations is that whose modulus is unity, called the **Naperian** system, the above rule becomes :

The differential of the Naperian logarithm of a variable is the differential of the variable divided by the variable.

Unless otherwise mentioned, by $\log x$ will hereafter be meant Naperian logarithm of x. The base of this system is represented by the letter e, and its value will be shown to be 2.718281.

29. The exponential function.

I. *When the base is constant.*

Let $y = a^x$. Then, in the system whose base is b,

$$\log_b y = x \log_b a.$$

Differentiating both members,

$$m \frac{dy}{y} = \log_b a\, dx,$$

whence
$$dy = \frac{a^x \log_b a\, dx}{m},$$

or *the differential of an exponential function whose base is constant is the function into the logarithm of the base into the differential of the exponent, divided by the modulus of the system.*

For the Naperian system, $m = 1$, and we have

$$dy = a^x \log a\, dx.$$

If the exponential base is also the base of the logarithmic system, $\log a = 1$, and

$$dy = a^x dx,$$

or, e being the Naperian base, the differential of $y = e^x$ is

$$dy = e^x dx.$$

II. *When the base is variable.*

Let $y = x^x$. Then $\log y = z \log x$. Differentiating both members,

$$\frac{dy}{y} = z\frac{dx}{x} + \log x dz,$$

whence $dy = x^z z\dfrac{dx}{x} + x^z \log x dz = zx^{z-1} dx + x^z \log x dz,$

or *the differential of an exponential function whose base is variable is the sum of the results obtained by differentiating first as if the exponent were constant and then as if the base were constant.*

If $z = x$, $y = x^x$, and $dy = x^x (1 + \log x) dx$.

EXAMPLES. Differentiate:

1. $y = \log (3ax + x^3)$. $dy = 3\,\dfrac{a + x^2}{3ax + x^3}dx.$

2. $y = \log x^2$. $dy = \dfrac{2\,dx}{x}.$

3. $y = (\log x)^2$. $dy = 2\log x\dfrac{dx}{x}.$

4. $y = \log (\log x)$. $dy = \dfrac{dx}{x\log x}.$

5. $y = x\log x$. $dy = (\log x + 1)\,dx.$

6. $y = \dfrac{1}{\log x^2}$. $dy = -\dfrac{2\,dx}{x\,(\log x^2)^2}.$

7. $y = \log (1 + x^2)^2$, or $2\log (1 + x^2)$.

$$dy = \frac{4\,x dx}{1 + x^2}.$$

8. $y = \log \sqrt{1+x^2}.$ $dy = \dfrac{x\,dx}{1+x^2}.$

9. $y = \log(\sqrt{1+x^2} + \sqrt{1-x^2}).$
$$dy = \frac{1}{x}\left(1 - \frac{1}{\sqrt{1-x^4}}\right)dx.$$

10. $y = \log\dfrac{1+\sqrt{x}}{1-\sqrt{x}},$ or $\log(1+\sqrt{x}) - \log(1-\sqrt{x}).$
$$dy = \frac{dx}{\sqrt{x}\,(1-x)}.$$

11. $y = \log[\sqrt{1-x}\,(1+x)],$ or $\frac{1}{2}\log(1-x) + \log(1+x).$
$$dy = \frac{(1-3x)\,dx}{2(1-x^2)}.$$

12. $y = \log_a 4\,x^{\frac{1}{4}}.$ $dy = \dfrac{m\,dx}{4\,x}.$

13. $y = e^x(1-x^2).$ $dy = e^x(1 - 2x - x^2)\,dx.$

14. $y = x^x.$ $dy = x^x(\log x + 1)\,dx.$

15. $y = x e^x.$ $dy = x^{e^x} e^x \dfrac{1 + x\log x}{x}\,dx.$

16. $y = e^{x^x}.$ $dy = e^{x^x} x^x(\log x + 1)\,dx.$

17. $y = x^{x^x}.$ $dy = x^{x^x} x^x\left[\log x\,(\log x + 1) + \dfrac{1}{x}\right]dx.$

18. $y = x^{\log x}.$ $dy = 2\,x^{\log x} \log x \,\dfrac{dx}{x}.$

19. $y = (\log x)^x.$ $dy = (\log x)^x\left[\log(\log x) + \dfrac{1}{\log x}\right]dx.$

20. $y = \log(e^x - e^{-x}).$ $dy = \dfrac{e^x + e^{-x}}{e^x - e^{-x}}\,dx.$

21. $y = \dfrac{e^x + e^{-x}}{e^x - e^{-x}}.$ $dy = -\dfrac{4\,dx}{(e^x - e^{-x})^2}.$

22. $y = \left(\dfrac{n}{x}\right)^x.$ $dy = \left(\dfrac{n}{x}\right)^x\left(\log\dfrac{n}{x} - 1\right)dx.$ •

23. $y = \left(\dfrac{x}{n}\right)^x$. $\qquad\qquad dy = \left(\dfrac{x}{n}\right)^x\left(\log\dfrac{x}{n} + 1\right)dx.$

Algebraic functions may sometimes be differentiated with greater facility by first passing to logarithms, but it is usually more expeditious to differentiate directly. Differentiate the following by passing to logarithms.

24. $y = x\sqrt{1-x}(1+x)$.

$$\log y = \log x + \tfrac{1}{2}\log(1-x) + \log(1+x),$$

$$\frac{dy}{y} = \left(\frac{1}{x} - \frac{1}{2(1-x)} + \frac{1}{1+x}\right)dx\,;$$

$$\therefore\ dy = x\sqrt{1-x}(1+x)\frac{2+x-5x^2}{2x(1-x)(1+x)}\,dx$$

$$= \frac{2+x-5x^2}{2\sqrt{1-x}}dx.$$

25. $y = \dfrac{1+x}{1-x}$. $\qquad\qquad dy = \dfrac{2\,dx}{1-x^2}.$

26. $y = \dfrac{x(1+x^2)}{\sqrt{1-x^2}}$. $\qquad dy = \dfrac{1+3x^2-2x^4}{(1-x^2)^{\frac{3}{2}}}dx.$

27. $y = a^{b^x}$. $\qquad\qquad dy = a^{b^x}b^x\log a\log b\,dx.$

28. $y = x^{\frac{1}{x}}$. $\qquad\qquad dy = \dfrac{x^{\frac{1}{x}}(1-\log x)}{x^2}\,dx.$

29. $y = \log\dfrac{\sqrt{1+x}+\sqrt{1-x}}{\sqrt{1+x}-\sqrt{1-x}}$. $\qquad dy = -\dfrac{dx}{x\sqrt{1-x^2}}.$

30. $y = a^{\log x}$. $\qquad\qquad dy = a^{\log x}\log a\dfrac{dx}{x}.$

31. $y = \log(\sqrt{x-a}+\sqrt{x-b})$. $\quad dy = \dfrac{dx}{2\sqrt{(x-a)(x-b)}}.$

32. $y = \log(x-\sqrt{x^2-a^2})$. $\qquad dy = -\dfrac{dx}{\sqrt{x^2-a^2}}.$

33. $y = \dfrac{x}{e^x-1}$. $\qquad\qquad dy = \dfrac{e^x(1-x)-1}{(e^x-1)^2}dx.$

30. Applications. 1. Compare the rates of change of a number and its logarithm.

$x = \log_a y$, whence $\dfrac{dx}{dy} = \dfrac{m}{y}$, or the logarithm ($x$) changes faster or more slowly than the number (y), according as the number is less or greater than the modulus of the system. Since $m = 1$ in the Naperian system, the Naperian logarithms of proper fractions change faster than the fractions.

2. Compare the rates of change of a number and its logarithm in the common system, where the number is 534. The modulus of the system where base is 10 will be shown to be .434294, $\therefore \dfrac{m}{y} = \dfrac{.434294}{534} = .00081$, which will be found by examination of the tables to be the tabular difference corresponding to the number 534. Since $\dfrac{m}{y}$ changes with y, the relative rate of change of a number and its logarithm varies with the number. If we assume that for an increase of say .1 in the number there will be a proportional increase in the logarithm, the quantity to be added to the logarithm of 534 to obtain the logarithm of 534.1 will be .1 × .00081 = .000081. This, in fact, is the manner of using the tabular difference of the tables, and is equivalent to the supposition that $\dfrac{m}{y}$ remains constant while the number 534 changes to 534.1, a supposition which, although not strictly true, gives results sufficiently accurate within the limits of practice.

3. Find the tabular difference corresponding to the number 3217. *Ans.* .000135.

4. Prove that the rule for the differentiation of a power applies when the exponent is incommensurable.

Let $y = x^n$, n being incommensurable. Passing to logarithms (first squaring, as y may be negative, and negative numbers have no logarithms), $\log y = n \log x$, $\therefore \dfrac{dy}{y} = n \dfrac{dx}{x}$, or $dy = \pm nx^{n-1}dx$.

5. Prove in the same manner that the rule applies when the exponent is imaginary.

6. Find the slope of the logarithmic curve at the point where it crosses the axis of Y.

$x = \log_a y$, $\therefore \dfrac{dy}{dx} = \dfrac{y}{m}$, which for $x = 0$ (whence $y = 1$) becomes $\dfrac{1}{m}$. Since $y = 1$ when $x = 0$, whatever the base, the slopes of all logarithmic curves at their common point on the axis of Y vary inversely as the moduli of the systems. In the Naperian system $m = 1$, hence the slope of $x = \log y$ is the ordinate of the point of contact.

7. Find the equation of the tangent to $x = \log y$.

Ans. $y - y' = y'(x - x')$.

8. Show that the subtangent of $x = \log_a y$ is constant and equal to the modulus of the system. Also find the subnormal.

Ans. $y' \dfrac{dy'}{dx'} = m$; $y' \dfrac{dx'}{dy'} = \dfrac{y'^2}{m}$.

9. Compare the rates of change of x and its xth power when $x = 1$. *Ans.* The rates are equal.

10. Compare the rates of change of x and its xth root when $x = e$.

Ans. $\dfrac{dy}{dx} = 0$.

The Trigonometric Functions.

31. Circular measure of an angle.

Any angle AOB, measured in degrees, may also be measured by the ratio of its arc to the radius of its arc, since for any given angle this ratio is constant whatever the radius of the arc. If the arc θ be described with a radius equal to the linear unit, then, since $x = r\theta$ (Fig. 6), $\dfrac{x}{r} = \theta$, or, by this method, the angle is measured by the arc intercepted at a unit's distance. To express the angle

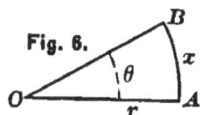
Fig. 6.

$n°$ in circular measure, we have $\dfrac{x}{r} = \dfrac{2\pi r}{r} = 2\pi$ for the circular measure of $360°$; hence the circular measure of $1°$ is $\dfrac{2\pi}{360} = \dfrac{\pi}{180}$, and of $n°$ is $\dfrac{n\pi}{180}$; or the circular measure of an angle is expressed by multiplying the number of degrees by $\dfrac{\pi}{180}$.

Since $\dfrac{x}{r} = 1$ when $x = r$, the unit of circular measure is the angle whose arc equals its radius; or, making $\dfrac{n\pi}{180} = 1$, $n = \dfrac{180}{\pi}$ $= 57°.3$ nearly.

32. Differential of sin x.

Let the point P move in the circular path AB, x being the length of the path, estimated from A, at any instant when the generating point is at P. Then

$$PD = y = \sin x.$$

If at this instant the motion of P should become uniform along the tangent at P, the changes in AP and PD would also become uniform. Hence if PQ, RQ, are what the increments of x and y would be in any interval dt, $PQ = dx$ and $RQ = dy = d(\sin x)$. But $RQ = PQ \cos AOP$. Hence $dy = \cos x\, dx$, or *the differential of the sine of an angle is the cosine of the angle into the differential of the angle.*

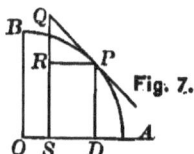

33. Differential of cos x.

In Fig. 7, $SD = RP$, being the decrement of OD simultaneous with RQ and PQ, is the differential of $\cos x$. Hence, if $OD = y = \cos x$, $dy = RP = -PQ \sin AOP = -\sin x\, dx$. Otherwise: $y = \cos x = \sqrt{1 - \sin^2 x}$, whence

$$dy = \frac{-2\sin x\, d(\sin x)}{2\sqrt{1 - \sin^2 x}} = -\frac{\sin x \cos x\, dx}{\cos x} = -\sin x\, dx,$$

or *the differential of the cosine of an angle is minus the sine of the angle into the differential of the angle.*

34. Differential of tan x.

Let $y = \tan x = \dfrac{\sin x}{\cos x}$. Then

$$dy = \frac{\cos x\, d(\sin x) - \sin x\, d(\cos x)}{\cos^2 x} = \frac{\cos^2 x + \sin^2 x}{\cos^2 x}dx$$

$$= \frac{dx}{\cos^2 x} = \sec^2 x\, dx,$$

or *the differential of the tangent of an angle is the square of the secant of the angle into the differential of the angle.*

35. Differential of cot x.

Let $y = \cot x = \tan\left(\dfrac{\pi}{2} - x\right)$. Then

$$dy = \sec^2\left(\frac{\pi}{2} - x\right)(-dx) = -\operatorname{cosec}^2 x\, dx,$$ a result which may also be obtained by differentiating $y = \cot x = \dfrac{\cos x}{\sin x}$. Hence

The differential of the cotangent of an angle is minus the square of the cosecant of the angle into the differential of the angle.

36. Differential of sec x.

Let $y = \sec x = \dfrac{1}{\cos x}$. Then

$$dy = -\frac{d(\cos x)}{\cos^2 x} = \frac{\sin x\, dx}{\cos^2 x} = \sec x \tan x\, dx,$$

or *the differential of the secant of an angle is the secant of the angle into the tangent of the angle into the differential of the angle.*

37. Differential of cosec x.

Let $y = \operatorname{cosec} x = \sec\left(\dfrac{\pi}{2} - x\right)$. Then

$$dy = \sec\left(\frac{\pi}{2} - x\right)\tan\left(\frac{\pi}{2} - x\right)(-dx) = -\operatorname{cosec} x \cot x\, dx,$$

or *the differential of the cosecant of an angle is minus the cosecant of the angle into the cotangent of the angle into the differential of the angle.*

38. Differential of vers x.

Let $y = \text{vers } x = 1 - \cos x$. Then $dy = \sin x\,dx$,

or *the differential of the versine of an angle is the sine of the angle into the differential of the angle.*

39. Differential of covers x.

Let $y = \text{covers } x = 1 - \sin x$. Then $dy = -\cos x\,dx$,

or *the differential of the coversine of an angle is minus the cosine of the angle into the differential of the angle.*

EXAMPLES.   Differentiate:

1. $y = \sin 6x$. $dy = 6 \cos 6x\,dx$.

2. $y = \cos x^2$. $dy = -2x \sin x^2\,dx$.

3. $y = \cos^2 x$. $dy = -\sin 2x\,dx$.

4. $y = \tan (3 - 5x^2)^2$. $dy = -20x(3 - 5x^2)\sec^2(3 - 5x^2)^2\,dx$.

5. $y = \sin^2 (1 - 2x^2)^2$. $dy = -8x(1 - 2x^2)\sin 2(1 - 2x^2)^2\,dx$.

6. $y = (\sin x \cos x)^2$. $dy = \sin 2x \cos 2x\,dx$.

7. $y = \sin 2x \cos 2x$. $dy = 2 \cos 4x\,dx$.

8. $y = \sin^2 (1 - x^2)^2$. $dy = -8x (1 - x^2) \sin (1 - x^2)^2\,dx$.

9. $y = \tan x + \sec x$. $dy = \dfrac{1 + \sin x}{\cos^2 x}\,dx$.

10. $y = x + \sin x \cos x$. $dy = 2 \cos^2 x\,dx$.

11. $y = \tan \sqrt{1 - x^2}$. $dy = -\sec^2 \sqrt{1 - x^2}\,\dfrac{x\,dx}{\sqrt{1 - x^2}}$.

12. $y = \sin (\log x)$. $dy = \dfrac{\cos (\log x)}{x}\,dx$.

13. $y = \log (\cot x)$. $dy = -\dfrac{2\,dx}{\sin 2x}$.

14. $y = m \sin^n ax$. $dy = amn \sin^{n-1} ax \cos ax\,dx$.

15. $y = \sin^x x$. $dy = \sin^x x(\log \sin x + x \cot x)\,dx$.

16. $y = \text{vers}\, \dfrac{x}{a}.$ $\qquad dy = \dfrac{1}{a} \sin \dfrac{x}{a}\, dx.$

17. $y = \sin e^x.$ $\qquad dy = e^x \cos e^x\, dx.$

18. $y = x^2 \cos x^2.$ $\qquad dy = 2x(\cos x^2 - x^2 \sin x^2)\, dx.$

19. $y = \sin \dfrac{a}{x}.$ $\qquad dy = -\dfrac{a}{x^2} \cos \dfrac{d}{x}\, dx.$

20. $y = \log(\sin x).$ $\qquad dy = \cot x\, dx.$

21. $y = \sin ax \sin^a x.$ $\qquad dy = a \sin^{a-1} x \sin(ax + x)\, dx.$

22. $y = \tan a^{\frac{1}{x}}.$ $\qquad dy = -\dfrac{\sec^2 a^{\frac{1}{x}} \log a \cdot a^{\frac{1}{x}} dx}{x^2}.$

23. $y = x^{\sin x}.$ $\qquad dy = x^{\sin x}\left(\dfrac{\sin x}{x} + \log x \cos x\right) dx.$

24. $y = (\sin x)^{\cos x}.$
$$dy = (\sin x)^{\cos x}\{\cot x \cos x - \sin x \log \sin x\}\, dx.$$

25. $y = \dfrac{(\sin nx)^m}{(\cos mx)^n}.$ $\qquad dy = \dfrac{mn(\sin nx)^{m-1}\cos(mx - nx)}{(\cos mx)^{n+1}}\, dx.$

The Circular Functions.

40. Differential of $\sin^{-1} x$.

Let $y = \sin^{-1} x$. Then $x = \sin y$. But $dx = \cos y\, dy$, hence

$$dy = \frac{dx}{\cos y} = \frac{dx}{\sqrt{1 - \sin^2 y}} = \frac{dx}{\sqrt{1 - x^2}},$$

or *the differential of an arc in terms of its sine is the differential of the sine, divided by the square root of 1 minus the square of the sine.*

41. Differential of $\cos^{-1} x$.

Let $y = \cos^{-1} x$. Then $x = \cos y$. But $dx = -\sin y\, dy$, hence

$$dy = -\frac{dx}{\sin y} = -\frac{dx}{\sqrt{1 - \cos^2 y}} = -\frac{dx}{\sqrt{1 - x^2}},$$

or *the differential of an arc in terms of its cosine is minus the differential of the cosine, divided by the square root of 1 minus the square of the cosine.*

42. Differential of $\tan^{-1}x$.

Let $y = \tan^{-1}x$. Then $x = \tan y$. But $dx = \sec^2 y\,dy$, hence

$$dy = \frac{dx}{\sec^2 y} = \frac{dx}{1 + \tan^2 y} = \frac{dx}{1 + x^2},$$

or *the differential of an arc in terms of its tangent is the differential of the tangent, divided by 1 plus the square of the tangent.*

43. Differential of $\cot^{-1}x$.

Let $y = \cot^{-1}x$. Then $x = \cot y$. But $dx = -\operatorname{cosec}^2 y\,dy$, hence

$$dy = \frac{-dx}{\operatorname{cosec}^2 y} = -\frac{dx}{1 + \cot^2 y} = -\frac{dx}{1 + x^2},$$

or *the differential of an arc in terms of its cotangent is minus the differential of the cotangent, divided by 1 plus the square of the cotangent.*

44. Differential of $\sec^{-1}x$.

Let $y = \sec^{-1}x$. Then $x = \sec y$. But $dx = \sec y \tan y\,dy$, hence

$$dy = \frac{dx}{\sec y \tan y} = \frac{dx}{\sec y \sqrt{\sec^2 y - 1}} = \frac{dx}{x\sqrt{x^2 - 1}},$$

or *the differential of an arc in terms of its secant is the differential of the secant, divided by the secant into the square root of the square of the secant minus 1.*

45. Differential of $\operatorname{cosec}^{-1}x$.

Let $y = \operatorname{cosec}^{-1}x$. Then $x = \operatorname{cosec} y$. But $dx = -\operatorname{cosec} x \cot x\,dx$, hence

$$dy = -\frac{dx}{x\sqrt{x^2 - 1}},$$

or *the differential of an arc in terms of its cosecant is minus the differential of the cosecant, divided by the cosecant into the square root of the square of the cosecant minus 1.*

46. Differential of $\mathrm{vers}^{-1}x$.

Let $y = \mathrm{vers}^{-1}x$. Then $x = \mathrm{vers}\, y$. But $dx = \sin y dy$, hence

$$dy = \frac{dx}{\sin y} = \frac{dx}{\sqrt{1 - \cos^2 y}} = \frac{dx}{\sqrt{1 - (1 - \mathrm{vers}\, y)^2}}$$

$$= \frac{dx}{\sqrt{1 - (1 - x)^2}} = \frac{dx}{\sqrt{2x - x^2}},$$

or *the differential of an arc in terms of its versine is the differential of the versine, divided by the square root of twice the versine minus the square of the versine.*

47. Differential of $\mathrm{covers}^{-1}x$.

Let $y = \mathrm{covers}^{-1}x$. Then $x = \mathrm{covers}\, y$. But $dx = -\cos y dy$, hence

$$dy = -\frac{dx}{\sqrt{2x - x^2}},$$

or *the differential of an arc in terms of its coversine is minus the differential of the coversine, divided by the square root of twice the coversine minus the square of the coversine.*

Examples. Differentiate:

1. $y = \sin^{-1} 2x^2.$ $\qquad dy = \dfrac{4xdx}{\sqrt{1 - 4x^4}}.$

2. $y = \cos^{-1}\sqrt{1 - x^2}.$ $\qquad dy = \dfrac{dx}{\sqrt{1 - x^2}}.$

3. $y = \sin^{-1}\dfrac{1 - x^2}{1 + x^2}.$ $\qquad dy = -\dfrac{2\,dx}{1 + x^2}.$

4. $y = \tan^{-1}a^{\frac{1}{z}}.$ $\qquad dy = -\dfrac{a^{\frac{1}{z}}\log a dx}{x^2\left(1 + a^{\frac{2}{z}}\right)}.$

5. $y = \tan^{-1} e^x$.

$$dy = \frac{dx}{e^x + e^{-x}}.$$

6. $y = \sin^{-1}(\tan x)$.

$$dy = \frac{\sec^2 x\, dx}{\sqrt{1 - \tan^2 x}}.$$

7. $y = \cos^{-1}(2 \cos x)$.

$$dy = -\frac{2 \sin x\, dx}{\sqrt{1 - 4 \cos^2 x}}.$$

8. $y = \cos^{-1}(\log x)$.

$$dy = -\frac{dx}{x\sqrt{1 - \log^2 x}}.$$

9. $y = \log(\cos^{-1} x)$.

$$dy = -\frac{dx}{\cos^{-1} x\sqrt{1 - x^2}}.$$

10. $y = \tan^{-1}\dfrac{2x}{1 + x^2}$.

$$dy = \frac{2(1 - x^2)\, dx}{1 + 6x^2 + x^4}.$$

11. $y = x \sin^{-1} x - \sqrt{1 - x^2}$.

$$dy = \left(\sin^{-1} x + \frac{2x}{\sqrt{1 - x^2}} \right) dx.$$

12. $x = r \operatorname{versin}^{-1}\dfrac{y}{r} - \sqrt{2ry - y^2}$.

$$dx = \frac{y\, dy}{\sqrt{2ry - y^2}}.$$

13. $y = (\sin^{-1} x)^z$.

$$dy = (\sin^{-1} x)^{z-1} \left\{ \frac{\sin^{-1} x \log(\sin^{-1} x) \sqrt{1 - x^2} + x}{\sqrt{1 - x^2}} \right\} dx.$$

14. $y = x^{\sin^{-1} x}$.

$$dy = x^{\sin^{-1} x} \left\{ \frac{\sin^{-1} x}{x} + \frac{\log x}{\sqrt{1 - x^2}} \right\} dx.$$

15. $y = \sin^{-1}\dfrac{x}{r}$.

$$dy = \frac{dx}{\sqrt{r^2 - x^2}}.$$

16. $y = \cos^{-1}\dfrac{x}{r}$.

$$dy = -\frac{dx}{\sqrt{r^2 - x^2}}.$$

17. $y = \tan^{-1}\dfrac{x}{r}$.

$$dy = \frac{r\, dx}{r^2 + x^2}.$$

18. $y = \cot^{-1}\dfrac{x}{r}$.

$$dy = -\frac{r\, dx}{r^2 + x^2}.$$

19. $y = \sec^{-1}\dfrac{x}{r}.$ $\qquad dy = \dfrac{r\,dx}{x\sqrt{x^2 - r^2}}.$

20. $y = \operatorname{cosec}^{-1}\dfrac{x}{r}.$ $\qquad dy = -\dfrac{r\,dx}{x\sqrt{x^2 - r^2}}.$

21. $y = \operatorname{vers}^{-1}\dfrac{x}{r}.$ $\qquad dy = \dfrac{dx}{\sqrt{2rx - x^2}}.$

22. $y = \operatorname{covers}^{-1}\dfrac{x}{r}.$ $\qquad dy = -\dfrac{dx}{\sqrt{2rx - x^2}}.$

23. $y = \tan^{-1}\sqrt{\dfrac{1 - \cos x}{1 + \cos x}}.$ $\qquad dy = \tfrac{1}{2}dx.$

24. $y = \sin^{-1}\sqrt{\sin x}.$ $\qquad dy = \tfrac{1}{2}\sqrt{1 + \operatorname{cosec} x}\, dx.$

25. $y = \log\left(\dfrac{1+x}{1-x}\right)^{\frac{1}{4}} + \tfrac{1}{2}\tan^{-1}x.$ $\quad dy = \dfrac{dx}{1 - x^4}.$

26. $y = \sec^{-1}\dfrac{1}{\sqrt{1 - x^2}}.$ $\qquad dy = \dfrac{dx}{\sqrt{1 - x^2}}.$

27. $y = \sin^{-1}\dfrac{x + 1}{\sqrt{2}}.$ $\qquad dy = \dfrac{dx}{\sqrt{1 - 2x - x^2}}.$

28. $y = (r^2 + x^2)\tan^{-1}\dfrac{x}{r}.$ $\qquad dy = \left(2x\tan^{-1}\dfrac{x}{r} + r\right)dx.$

29. $y = \tan^{-1}(n \tan x).$ $\qquad dy = \dfrac{n\,dx}{\cos^2 x + n^2 \sin^2 x}.$

30. $y = \cos^{-1}(\cos 2x).$ $\qquad dy = -2dx.$

48. Applications. 1. A wheel revolves about a fixed axis through its centre. Compare the velocity of a point on the rim with its velocity in a horizontal direction.

The horizontal velocity is evidently the rate of change of the cosine of the arc described by the point; hence, if the arc be denoted by x, $y = \cos x$, whence $dy = -\sin x\,dx$, which is also the relation between the rates of y and x. The point is therefore moving in a horizontal direction $\sin x$ times as fast as it is moving in the arc. At the highest point, where $x = 90°$, $\sin x = 1$, and $dy = -dx$, the rates being equal. At $x = 30°$, $\sin x = \frac{1}{2}$, and at this point the horizontal velocity is one-half that in the arc.

2. Compare the horizontal and vertical velocities of a point on the rim of a wheel which rolls without sliding with a constant velocity m on a horizontal line.

In this case the path of a point on the rim is a cycloid whose equation is $x = r \operatorname{vers}^{-1} \dfrac{y}{r} - \sqrt{2 ry - y^2}$,

whence $\dfrac{dx}{dt} = \dfrac{y}{\sqrt{2 ry - y^2}} \dfrac{dy}{dt}$ (1).

Fig. 8.

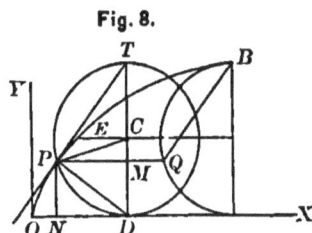

Since the wheel has a constant velocity m in a horizontal direction, and its centre C is always vertically over D, this velocity is the rate of change of

$$OD = r \operatorname{vers}^{-1} \frac{y}{r}.$$

Hence

$$\frac{d\left[r \operatorname{vers}^{-1} \dfrac{y}{r} \right]}{dt} = \frac{r}{\sqrt{2 ry - y^2}} \frac{dy}{dt} = m,$$

$$\therefore \frac{dy}{dt} = \frac{m}{r} \sqrt{2 ry - y^2}.$$

Substituting this value in (1),

$$\frac{dx}{dt} = \frac{y}{r} m.$$

Hence

$$\frac{ds}{dt} = \sqrt{\left(\frac{dx}{dt}\right)^2 + \left(\frac{dy}{dt}\right)^2} = m \sqrt{\frac{2y}{r}}.$$

At O, $y = 0$, and $\dfrac{dx}{dt} = \dfrac{dy}{dt} = \dfrac{ds}{dt} = 0$.

At B, $y = 2r$, and $\dfrac{dx}{dt} = \dfrac{ds}{dt} = 2m$, $\dfrac{dy}{dt} = 0$.

At E, $y = r$, and $\dfrac{dx}{dt} = \dfrac{dy}{dt} = m$, $\dfrac{ds}{dt} = m\sqrt{2}$.

3. Find the subnormal of the cycloid.

$$y' \frac{dy'}{dx'} = y' \frac{\sqrt{2ry' - y'^2}}{y'} = \sqrt{2ry' - y'^2}. \text{ But (Fig. 8) } \sqrt{2ry' - y'^2}$$

$= PM = ND$, or the normal passes through the foot of the vertical diameter of the circle when in position for the point P. Hence, also, the tangent passes through the upper extremity T. Therefore to draw a tangent and a normal at any point P, put the generating circle in position and join P with the extremities of its vertical diameter. Also, to draw a tangent parallel to a given line, draw BQ parallel to the given line, and PQ parallel to the base. Then P is the required point of tangency.

4. A man walks in a direction AB. Compare the rate of change of his distance from a point O with the rate of his angular motion about O.

Let fall the perpendicular $OD = p$ upon AB, and take O for the pole, OD for the polar axis. Then the equation of AB is

$$r = \frac{p}{\cos \theta}, \text{ whence } \frac{dr}{dt} = \frac{p \sin \theta}{\cos^2 \theta} \frac{d\theta}{dt}. \text{ For } \theta = 0, \frac{dr}{dt} = 0; \text{ for }$$

$\theta = 90°, \dfrac{dr}{dt} = \infty.$

5. An elliptical cam making two revolutions a second about a horizontal axis through one focus, gives motion to a bar in a vertical direction through the centre of revolution. The transverse axis being 6 and the eccentricity $\frac{2}{3}$, find the velocity of the bar when the angle between the vertical and the transverse axis is 60°; 90°.

$$r = \frac{a(1 - e^2)}{1 - e \cos \theta}, \text{ whence } \frac{dr}{dt} = -\frac{a(1 - e^2)e \sin \theta}{(1 - e \cos \theta)^2} \frac{d\theta}{dt}, \text{ which for}$$

$a = 3$, $e = \frac{2}{3}$, and $\dfrac{d\theta}{dt} = 4\pi$, becomes $-\dfrac{40 \sin \theta}{(3 - 2 \cos \theta)^2} \pi.$ When

$\theta = 60°, \dfrac{dr}{dt} = -5\sqrt{3}\,\pi;$ when $\theta = 90°, \dfrac{dr}{dt} = -\dfrac{40}{9}\pi.$

6. The crank of a steam engine is one foot in length and makes two revolutions a second. If the connecting rod is 5

feet in length, find the velocity of the piston when the crank makes angles of 45°, 135°, 90°, with the line of motion of the piston rod. Let a, b, x, represent the crank, connecting rod, and variable side of the triangle, respectively, and θ the angle between a and x. Then $x = a \cos \theta + \sqrt{b^2 - a^2 \sin^2 \theta}$, whence

$$\frac{dx}{dt} = - \left\{ a \sin \theta + \frac{a^2 \sin \theta \cos \theta}{\sqrt{b^2 - a^2 \sin^2 \theta}} \right\} \frac{d\theta}{dt},$$

which for $a = 1$, $b = 5$, $\dfrac{d\theta}{dt} = 4\pi$, becomes

$$- \left\{ \sin \theta + \frac{\sin \theta \cos \theta}{\sqrt{25 - \sin^2 \theta}} \right\} 4\pi.$$

Ans. $-\frac{16}{7} \sqrt{2}\,\pi$; $-\frac{12}{7} \sqrt{2}\,\pi$; -4π.

7. Find the slope of $y = \sin x$ at the points where the curve crosses X. *Ans.* ± 1.

8. Find the angle at which $y = \sin x$ crosses $y = \cos x$. *Ans.* $\tan^{-1} 2\sqrt{2}$.

9. Find the length of the normal to the cycloid. *Ans.* $\sqrt{2ry'}$.

CHAPTER III.

SUCCESSIVE DIFFERENTIATION.

49. Equicrescent variable. *A variable which changes uniformly*, that is, *whose rate is constant*, is said to be **equicrescent**.

50. *The differential of an equicrescent variable is constant.*

For, if x be equicrescent, its rate $\dfrac{dx}{dt}$ is constant. But dt is constant; hence dx is also constant.

It is evident that, if $\dfrac{dx}{dt}$ is not constant, dx is a variable.

The above is a direct consequence of the definitions; for the differential of a variable is what would be its change during any interval were its rate of change to remain throughout the interval what it was at its beginning. If the rate varies from instant to instant, differentials corresponding to equal intervals also vary; while if the rate remains the same, these differentials are equal.

51. Successive derived equations.

Let $y = f(x)$. Then $dy = f'(x)\,dx$, in which $f'(x) = \dfrac{dy}{dx}$, the first derivative.

Now, in general, dy, or $f'(x)\,dx$, is a variable. For dx is a variable unless x is equicrescent; and $f'(x)$ is a variable unless $f(x)$ is linear, in which case it can be reduced to the form $y = mx + b$, whence $\dfrac{dy}{dx} = f'(x) = m$, a constant. Hence, unless the function is linear and x is equicrescent, $dy = f'(x)\,dx$ is variable, and, being true for all values of x, can be differentiated, thus forming a **second derived equation** which may in its turn be differentiated, a repetition of this process leading to

49

successive derived equations called the first, second, third, etc., in order.

Since differentiation introduces no function which has not been already treated, the successive derived equations are obtained by the rules already established.

52. Notation. The second differential of a variable x is represented by the symbol d^2x, read 'second differential of x,' the exponent being a symbol of operation indicating how many times the variable has been differentiated. The student will observe the different meanings of the forms d^2x, dx^2, and $d(x)^2$.

ILLUSTRATION. Given $y^2 = 2px$. The first derived equation is $2ydy = 2pdx$, or $ydy = pdx$. Differentiating again, we have

$$yd(dy) + dyd(y) = pd(dx),$$

or, in the above notation,

$$yd^2y + dy^2 = pd^2x,$$

which is the second derived equation. Differentiating again,

$$yd(d^2y) + d^2yd(y) + 2\,dyd(dy) = pd(d^2x),$$

or $\qquad yd^3y + d^2ydy + 2\,dyd^2y = pd^3x,$

whence $\qquad yd^3y + 3\,dyd^2y = pd^3x,$

which is the third derived equation.

If x were equicrescent, the successive derived equations would be much simplified. For when x is equicrescent, dx is constant, and, since the differential of a constant is zero, all the successive differentials of x after the first would vanish. Thus, in the above illustration, $d^2x = d^3x =$ etc. $= 0$, and the successive derived equations become

$$ydy = pdx,$$
$$yd^2y + dy^2 = 0,$$
$$yd^3y + 3\,dyd^2y = 0.$$

53. Remark. It is important to observe that in most cases it is permissible to consider the variable equicrescent and thus secure the simplicity above noted. For example, let $y = f(x)$ be the equation of any plane curve. The assumption that x is equicrescent, or that $\dfrac{dx}{dt}$ is constant, implies that the velocity of the generating point in the direction of the axis of X is constant. Now, so far as the geometrical properties of the curve are concerned, these being independent of the velocity of the generating point, we are at liberty to make *any* assumption regarding the velocity which will facilitate their investigation. We therefore assume the velocity-law in the curve such that the motion in the direction of the axis of X is uniform.

Again: suppose a right cylinder is inscribed in a right cone, the problem being to find, of all right cylinders so inscribed, that one whose volume is the greatest. If the radius of the base and altitude of the cone are b and a, and those of the cylinder x and y, we have

$$b : a :: x : a - y,$$

whence

$$x = \frac{b}{a}(a - y);$$

and if V is the volume of the cylinder,

$$V = \pi y x^2 = \pi \frac{b^2}{a^2} y (a - y)^2.$$

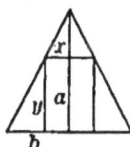

Fig. 9.

Now in determining the greatest value of V, it is evidently immaterial whether we regard y equicrescent or not, since the cylinder of greatest volume is independent of the law of change of y.

In functions of a single variable, unless mention is made to the contrary, the variable will hereafter be regarded equicrescent.

EXAMPLES. Regarding x equicrescent, form:

1. The second derived equations of

$$a^2y^2 + b^2x^2 = a^2b^2, \qquad a^2y\,d^2y + a^2dy^2 + b^2dx^2 = 0.$$
$$y^2 + x^2 = R^2, \qquad y\,d^2y + dy^2 + dx^2 = 0.$$
$$xy = m, \qquad 2\,dy\,dx + x\,d^2y = 0.$$
$$y = x^2 \log x, \qquad d^2y = 2 \log x\,dx + 3\,dx^2.$$

2. The fifth derived equation of $y = x^4 \log x$.

$$d^5y = \frac{24}{x}dx^5.$$

3. The fourth derived equation of $y = \dfrac{x^3}{1-x}$.

$$d^4y = \frac{24}{(1-x)^5}dx^4.$$

4. The third derived equations of:

$y = \tan x,$ $\qquad\qquad d^3y = 2(3\sec^2 x - 2)\sec^2 x dx^3.$

$y = e^x,$ $\qquad\qquad d^3y = e^x dx^3.$

$y = \dfrac{1}{x},$ $\qquad\qquad d^3y = -\dfrac{6}{x^4}dx^3.$

$y = \cos x,$ $\qquad\qquad d^3y = \sin x dx^3.$

5. Prove that the nth derived equation of $y = a^x$ is

$$d^ny = (\log a)^n a^x dx^n.$$

6. If $y = \log \sin x$, prove that $d^3y = 2\dfrac{\cot x}{\sin^2 x}dx^3.$

7. If $y = \sin^{-1}\sqrt{x}$, prove that $d^2y = -\dfrac{1-2x}{4(x-x^2)^{\frac{3}{2}}}dx^2.$

8. If $y = m\cos^m mx$, prove that

$$d^2y = -m^4\{(\cos mx)^m - (m-1)(\cos mx)^{m-2}\sin^2 mx\}dx^2.$$

9. If $y = ax^4$, show that $\qquad d^5y = 0.$

54. Successive derivatives, or differential coefficients.

Let $y = f(x)$, in which x is equicrescent. The first derivative of $f(x)$ has been defined as $\dfrac{dy}{dx} = \dfrac{d[f(x)]}{dx}$, and is the ratio of the rates of change of the function and its variable. Since the first derivative is variable except when $f(x)$ is linear, it is in general a function of x and may be denoted by $f'(x)$, or $\dfrac{dy}{dx} = f'(x)$; it may therefore be differentiated in its turn, and a second derivative formed by dividing $d[f'(x)]$ by dx, and this process may evidently be continued until a derivative is reached which is constant. The successive derivatives thus obtained are called in order the **first, second, third,** etc., **derivatives,** and are denoted by $f'(x)$, $f''(x)$, $f'''(x)$, etc.

Since each derivative is obtained from the preceding one in the same manner that $f'(x)$ is obtained from $f(x)$, it follows that:

1. *The nth derivative of $f(x)$ is the ratio of the rate of change of the $(n-1)th$ derivative to that of the variable.*

2. *The nth derivative of $f(x)$ may be obtained either by differentiating the $(n-1)th$ derivative and dividing by dx, or by dividing the nth derived equation by dx^n.*

ILLUSTRATION. Given $y = a + bx^3$. The first derivative is

$$\frac{dy}{dx} = 3bx^2 = f'(x).$$

Differentiating, remembering that dx is constant,

$$\frac{d^2y}{dx} = 6bxdx,$$

whence the second derivative

$$\frac{d^2y}{dx^2} = 6bx = f''(x).$$

Differentiating again,

$$\frac{d^3y}{dx^2} = 6bdx,$$

whence the third derivative

$$\frac{d^3y}{dx^3} = 6b = f'''(x).$$

Here the process ends, since the third derivative is constant. Otherwise, differentiating $y = a + bx^3$ successively three times, the successive derived equations are

$$dy = 3bx^2dx, \quad d^2y = 6bxdx^2, \quad d^3y = 6bdx^3,$$

and, dividing the last by dx^3,

$$\frac{d^3y}{dx^3} = 6\,b,$$

as before.

55. Sign of the nth derivative. Since $f'(x)$ is positive or negative as $f(x)$ is an increasing or decreasing function (Art. 22), and since $f''(x)$ is the first derivative of $f'(x)$, $f'''(x)$ the first derivative of $f''(x)$, etc., therefore $f^n(x)$ *will be positive or negative as $f^{n-1}(x)$ is an increasing or a decreasing function.*

EXAMPLES. 1. If $y = mx^m$, prove that $\dfrac{d^3y}{dx^3}$, or $f'''(x)$, is

$$m^2(m-1)(m-2)x^{m-3}.$$

$$f'(x) = m^2x^{m-1}.$$
$$f''(x) = m^2(m-1)x^{m-2}.$$
$$f'''(x) = m^2(m-1)(m-2)x^{m-3}.$$

2. If $y = e^x \sin x$, prove that $\dfrac{d^2y}{dx^2} = 2\,e^x \cos x$.

$$f'(x) = e^x \cos x + e^x \sin x = e^x(\cos x + \sin x).$$
$$f''(x) = e^x(-\sin x + \cos x) + e^x(\cos x + \sin x) = 2\,e^x \cos x.$$

3. If $y = \log \cos x$, prove that $f^{iv}(x) = -2\sec^2x\,(3\sec^2x - 2)$.

4. If $y = \sqrt{1-x^2}$, prove that $\dfrac{d^2y}{dx^2} = -\dfrac{1}{y^3}$.

5. If $y = e^{\sin x}$, prove that

$$f'''(x) = e^{\sin x}\cos x\,(\cos^2 x - 3\sin x - 1).$$

6. If $y^2 = \sec 2x$, prove that $f''(x) = 3\,y^5 - y$.

7. If $y = a^{bx}$, prove that $f^v(x) = b^5 \log^5 a \cdot a^{bx}$.

8. If $y^2 = 2px$, and y is equicrescent, prove that $\dfrac{d^2x}{dy^2} = \dfrac{1}{p}$.

The following first and second derivatives, being of frequent use hereafter, may be here established for future reference. In all implicit functions of two variables, x will be regarded as the equicrescent variable unless otherwise mentioned.

9. The ellipse, $a^2y^2 + b^2x^2 = a^2b^2$.

$$f'(x) = -\frac{b^2x}{a^2y} = \mp \frac{b}{a}\frac{x}{\sqrt{a^2-x^2}}. \qquad f''(x) = -\frac{b^4}{a^2y^3}.$$

10. The circle, $y^2 + x^2 = R^2$.

$$f'(x) = -\frac{x}{y} = \mp \frac{x}{\sqrt{R^2-x^2}}. \qquad f''(x) = -\frac{R^2}{y^3}.$$

11. The hyperbola, $a^2y^2 - b^2x^2 = -a^2b^2$.

$$f'(x) = \frac{b^2x}{a^2y} = \pm\frac{b}{a}\frac{x}{\sqrt{x^2-a^2}}. \qquad f''(x) = -\frac{b^4}{a^2y^3}.$$

12. The hyperbola, $xy = m$.

$$f'(x) = -\frac{y}{x} = -\frac{m}{x^2}. \qquad f''(x) = \frac{2y}{x^2} = \frac{2m}{x^3}.$$

13. The parabola, $y^2 = 2px$.

$$f'(x) = \frac{p}{y} = \pm\frac{p}{\sqrt{2px}}. \qquad f''(x) = -\frac{p^2}{y^3}.$$

14. The cubical parabola, $y^3 = a^2x$.

$$f'(x) = \frac{a^2}{3y^2} = \frac{a^{\frac{4}{3}}}{3x^{\frac{2}{3}}}. \qquad f''(x) = -\frac{2}{9}\frac{a^4}{y^5}.$$

15. The semi-cubical parabola, $ay^2 = x^3$.

$$f'(x) = \frac{3x^2}{2ay} = \pm\frac{3}{2}\sqrt{\frac{x}{a}}. \qquad f''(x) = \frac{3x^4}{4a^2y^3}.$$

16. The witch, $x^2y = 4a^2(2a-y)$.

$$f'(x) = -\frac{2xy}{x^2+4a^2} = -\frac{16a^2x}{(x^2+4a^2)^2}.$$

$$f''(x) = 2y\frac{3x^2-4a^2}{(x^2+4a^2)^2}.$$

17. The cycloid, $x = r\,\text{versin}^{-1}\frac{y}{r} - \sqrt{2ry-y^2}$.

$$f'(x) = \pm\frac{\sqrt{2ry-y^2}}{y}. \qquad f''(x) = -\frac{r}{y^2}.$$

18. The cissoid, $y^2 = \dfrac{x^3}{2a - x}$.

$$f'(x) = \pm x^{\frac{1}{2}} \frac{3a - x}{(2a - x)^{\frac{3}{2}}}. \qquad f''(x) = \pm \frac{3a^2}{x^{\frac{1}{2}}(2a - x)^{\frac{5}{2}}}.$$

19. The hypocycloid, $x^{\frac{2}{3}} + y^{\frac{2}{3}} = a^{\frac{2}{3}}$.

$$f'(x) = \frac{y^{\frac{1}{3}}}{x^{\frac{1}{3}}}. \qquad f''(x) = \frac{1}{3} \frac{a^{\frac{2}{3}}}{y^{\frac{1}{3}} x^{\frac{4}{3}}}.$$

20. The catenary, $y = \dfrac{a}{2}(e^{\frac{x}{a}} + e^{-\frac{x}{a}})$.

$$f'(x) = \frac{1}{2}(e^{\frac{x}{a}} - c^{-\frac{x}{a}}). \qquad f''(x) = \frac{y}{a^2}.$$

21. The logarithmic curve, $x = \log y$.

$$f'(x) = f''(x) = y.$$

22. The sinusoid, $y = \sin x$.

$$f'(x) = \cos x. \qquad f''(x) = -\sin x = -y.$$

56. Remark. If a function becomes infinite for a *finite value* of the variable, its derived functions also become infinite.

For if the function be an algebraic one, it can become infinite for a finite value of the variable only by having the form of a fraction whose denominator vanishes for that value, and, in differentiating to form the derived functions, this denominator never disappears. So that if $f(x) = \infty$ when $x = x'$, $f'(x)$, $f''(x)$, etc., also become infinity when $x = x'$. Examination of the transcendental functions leads to the same conclusion. Thus $\log x$ becomes infinity when $x = 0$, as do also all its derivatives $\dfrac{1}{x}$, $-\dfrac{1}{x^2}$, etc.; and $a^{\frac{1}{x}}$, $\tan x$, $\sec x$, illustrate the same fact.

This is not necessarily true when $f(x)$ becomes infinity for an *infinite value* of the variable. Thus, $\log x = \infty$ when $x = \infty$; but $f'(x) = \dfrac{1}{x}$ becomes zero for $x = \infty$.

57. Notation. To denote what a function becomes for a particular value of the variable, the variable is replaced by its particular value. Thus, $f(a)$, $f(0)$, $f(x')$, represent what $f(x)$ becomes when $x = a$, $x = 0$, $x = x'$, respectively. The particular value may also be written as a subscript in either of the following ways:

$$\frac{1}{x}\Big]_{x=\infty} = 0, \quad \frac{1}{x}\Big]_{\infty} = 0,$$

read '$\dfrac{1}{x}$ equals zero when x is infinity.'

58. Change of the equicrescent variable.

In forming the successive derivatives of $y = f(x)$ we have considered x equicrescent, that is, dx constant, and hence $d^2x = d^3x = $ etc. $= 0$.

If x is not equicrescent, dx is a variable, and

$$\frac{d\left(\dfrac{dy}{dx}\right)}{dx} = \frac{dx\,d^2y - dy\,d^2x}{dx^3}, \tag{1}$$

which is the general form of the second derivative when neither x nor y is equicrescent.

Differentiating (1), regarding dx and dy as variables, we have

$$\frac{d\left(\dfrac{dx\,d^2y - dy\,d^2x}{dx^3}\right)}{dx} = \frac{(d^3y\,dx - d^3x\,dy)\,dx - 3(d^2y\,dx - d^2x\,dy)\,d^2x}{dx^5}, \tag{2}$$

which is the general form of the third derivative when neither x nor y is equicrescent. The general forms of the third, fourth, etc., derivatives may be found in like manner.

If in (1) and (2) x is equicrescent, $d^2x = d^3x = 0$, and we have

$$\frac{d^2y}{dx^2} \quad \text{and} \quad \frac{d^3y}{dx^3}, \tag{3}$$

while if y is equicrescent, $d^2y = d^3y = 0$, and we have

$$-\frac{dy\,d^2x}{dx^3} \quad \text{and} \quad \frac{3(d^2x)^2dy - d^3x\,dy\,dx}{dx^5}. \tag{4}$$

Thus the forms of the successive derivatives, *after the first,* differ, according as the variable, the function, or neither, is considered equicrescent.

To transform a differential expression which has been formed on the hypothesis that x is equicrescent into its equivalent in which neither x nor y is equicrescent, we have only to replace the successive derivatives by the general forms (1), (2), etc.

To change the equicrescent variable from x to y, we replace the successive derivatives by (4) directly, or by the general forms, and then make $d^2y = d^3y = $ etc. $= 0$.

To transform a differential expression formed on the hypothesis that either x or y is equicrescent into its equivalent in terms of a new equicrescent variable θ, we first replace the successive derivatives by their general forms when neither x nor y is equicrescent, and then substitute for x, y, dy, dx, d^2y, d^2x, etc., their values in terms of θ.

EXAMPLES.   1. Change the equicrescent variable from x to y in the expression $y\dfrac{d^2y}{dx^2} + \dfrac{dy^2}{dx^2} + 1 = 0$.

Replacing $\dfrac{d^2y}{dx^2}$ by $-\dfrac{dyd^2x}{dx^3}$, we have $-y\dfrac{dyd^2x}{dx^3} + \dfrac{dy^2}{dx^2} + 1 = 0$, or, dividing by dy^2 and multiplying by dx^3,

$$y\frac{d^2x}{dy^2} - \frac{dx^3}{dy^3} - \frac{dx}{dy} = 0,$$

in which the position of dy indicates that y is the equicrescent variable.

2. Change the equicrescent variable from x to z in the equation $x^4\dfrac{d^2y}{dx^2} + a^2y = 0$, having given $x = \dfrac{1}{z}$.

Replacing $\dfrac{d^2y}{dx^2}$ by $\dfrac{dxd^2y - dyd^2x}{dx^3}$, we have, after substituting $dx = -\dfrac{dz}{z^2}$ and $d^2x = \dfrac{2\,dz^2}{z^3}$, $\dfrac{d^2y}{dz^2} + \dfrac{2}{z}\dfrac{dy}{dz} + a^2y = 0$.

3. Change the equicrescent variable from x to t in the equation $\dfrac{d^2y}{dx^2} + \dfrac{1}{x}\dfrac{dy}{dx} + y = 0$, having given $x^2 = 4t$.

From $x^2 = 4t$, $dx = \dfrac{dt}{\sqrt{t}}$, $d^2x = -\dfrac{dt^2}{2t\sqrt{t}}$. Hence, replacing $\dfrac{d^2y}{dx^2}$ by the general form as in Ex. 2, we find $t\dfrac{d^2y}{dt^2} + \dfrac{dy}{dt} + y = 0$.

4. Change the equicrescent variable from x to θ in

$$\frac{d^2y}{dx^2} - \frac{x}{1-x^2}\frac{dy}{dx} + \frac{y}{1-x^2} = 0,$$

having given $x = \sin \theta$.

$dx = \cos \theta\, d\theta$, $d^2x = -\sin \theta\, d\theta^2$, $1 - x^2 = 1 - \sin^2 \theta = \cos^2\theta$.

Hence

$$\frac{dx\,d^2y - dy\,d^2x}{dx^3} - \frac{x}{1-x^2}\frac{dy}{dx} + \frac{y}{1-x^2} = \frac{\cos\theta\, d\theta\, d^2y + \sin\theta\, d\theta^2\, dy}{\cos^3\theta\, d\theta^3}$$

$$-\frac{\sin\theta}{\cos^2\theta}\frac{dy}{\cos\theta\, d\theta} + \frac{y}{\cos^2\theta} = 0, \text{ or } \frac{d^2y}{d\theta^2} + y = 0.$$

5. Change the equicrescent variable from x to θ in the expression $\dfrac{\left\{1 + \dfrac{dy^2}{dx^2}\right\}^{\frac{3}{2}}}{\dfrac{d^2y}{dx^2}}$, having given $x = a\cos\theta$, $y = b\sin\theta$.

$$\text{Ans. } \frac{(a^2\sin^2\theta + b^2\cos^2\theta)^{\frac{3}{2}}}{ab}.$$

6. If $(a^2 - x^2)\dfrac{d^2z}{dx^2} - \dfrac{a^2}{x}\dfrac{dz}{dx} - z = 0$, show that $x^2\dfrac{d^2z}{dy^2} - z = 0$, having given $x^2 + y^2 = a^2$.

We have from $x^2 + y^2 = a^2$, $dx = -\dfrac{y}{x}dy$, $d^2x = -\dfrac{a^2dy^2}{x^3}$.

Replacing $\dfrac{d^2z}{dx^2}$ by $\dfrac{dx\,d^2z - dz\,d^2x}{dx^3}$, substituting the above values of dx and d^2x, and for $a^2 - x^2$ its equal y^2, the given expression becomes $x\dfrac{d^2z}{dy^2} - z = 0$.

7. Change the equicrescent variable from x to θ in the expression

$$\frac{\left(1 + \frac{dy^2}{dx^2}\right)^{\frac{3}{2}}}{\frac{d^2y}{dx^2}},$$

having given $y = r \sin \theta,\ x = r \cos \theta$.

$dy = \sin \theta dr + r \cos \theta d\theta,\quad dx = \cos \theta dr - r \sin \theta d\theta.$

$d^2y = \sin \theta d^2r + 2 \cos \theta d\theta dr - r \sin \theta d\theta^2.$

$d^2x = \cos \theta d^2r - 2 \sin \theta d\theta dr - r \cos \theta d\theta^2.$

Substituting these values, we find

$$\frac{\left(1 + \frac{dy^2}{dx^2}\right)^{\frac{3}{2}}}{\frac{d^2ydx - d^2xdy}{dx^3}} = \frac{(dx^2 + dy^2)^{\frac{3}{2}}}{d^2ydx - d^2xdy} = \frac{\left(\frac{dr^2}{d\theta^2} + r^2\right)^{\frac{3}{2}}}{2\frac{dr^2}{d\theta^2} - r\frac{d^2r}{d\theta^2} + r^2}.$$

APPLICATIONS OF SUCCESSIVE DIFFERENTIATION.

Accelerations.

59. Acceleration. *Signification of* $\frac{d^2s}{dt^2}$. Velocity has been defined (Art. 6) as the rate of change of the distance passed over by a moving point; hence if s be the distance and v the velocity,

$$v = \frac{ds}{dt}.$$

The rate of change of v is called the **acceleration**.
Now the rate of change of v is

$$\frac{dv}{dt} = \frac{d\left(\frac{ds}{dt}\right)}{dt} = \frac{d^2s}{dt^2};$$

hence $\frac{d^2s}{dt^2}$ measures the acceleration of the point in its path.
Being the rate of v, *the acceleration is the amount by which the*

velocity would change in a unit of time were the velocity-rate to become constant at the instant considered. Thus, if at any instant the acceleration is said to be 5, it is meant that at that instant the velocity is changing at the rate of (5 feet per second) per second, or (5 miles per hour) per hour, according as the second or hour is the unit of time, and the foot or mile the unit of distance.

Cor. 1. If $v = \dfrac{ds}{dt}$ is constant, $\dfrac{d^2s}{dt^2} = 0$, or in uniform motion there is no acceleration.

Cor. 2. Since $\dfrac{dx}{dt}, \dfrac{dy}{dt}$, are the velocities in the directions of the axes, $\dfrac{d^2x}{dt^2}, \dfrac{d^2y}{dt^2}$, are the corresponding accelerations.

60. Signs of the axial accelerations.

$\dfrac{d^2x}{dt^2}$ may be plus or minus, and the sign is interpreted as follows: When plus, the velocity $\dfrac{dx}{dt}$ is accelerated in the positive direction of X. Thus, suppose $\dfrac{dx}{dt}$ is negative, or the point moving in the negative direction of X; then if $\dfrac{d^2x}{dt^2}$ is positive, the velocity is being accelerated in the direction $+ X$, that is, it is algebraically increasing, although numerically diminishing, till the motion is reversed, after which it increases numerically. In other words, the $\pm$ signs of the accelerations $\dfrac{d^2x}{dt^2}$, $\dfrac{d^2y}{dt^2}$, must be interpreted as an *algebraic* increase or decrease of the corresponding velocity whether the latter be positive or negative.

Examples. 1. A point moves in the arc of the parabola $y^2 = 2px$ with a constant velocity m. Find the accelerations in the directions of the axis.

From the equation of the path we have

$$\frac{dy}{dt} = \frac{p}{y} \frac{dx}{dt}, \tag{1}$$

and, by condition,

$$m = \frac{ds}{dt} = \sqrt{\left(\frac{dx}{dt}\right)^2 + \left(\frac{dy}{dt}\right)^2}. \qquad \text{[Art. 25.}$$

Substituting in this the value of $\frac{dy}{dt}$ from (1), we have

$$m = \sqrt{1 + \frac{p^2}{y^2}}\frac{dx}{dt}, \quad \therefore \quad \frac{dx}{dt} = \frac{my}{\sqrt{y^2 + p^2}}, \qquad (2)$$

which in (1) gives

$$\frac{dy}{dt} = \frac{mp}{\sqrt{y^2 + p^2}}. \qquad (3)$$

Differentiating (2) and (3), and dividing by dt to obtain their rates, we have

$$\frac{d^2x}{dt^2} = \frac{mp^2}{(y^2 + p^2)^{\frac{3}{2}}}\frac{dy}{dt} = \frac{m^2 p^3}{(y^2 + p^2)^2},$$

$$\frac{d^2y}{dt^2} = -\frac{mpy}{(y^2 + p^2)^{\frac{3}{2}}}\frac{dy}{dt} = -\frac{m^2 p^2 y}{(y^2 + p^2)^2}.$$

Since $\frac{d^2x}{dt^2}$ is always positive, the velocity along X is always increasing algebraically. $\frac{d^2y}{dt^2}$ is negative in the first angle and positive in the fourth, hence the velocity along Y is decreasing algebraically in the first angle and increasing algebraically in the fourth. These remarks are true when the point describes the arc of the parabola in either direction.

At $y = p$, $\frac{d^2x}{dt^2} = \frac{m^2}{4p}$, $\frac{d^2y}{dt^2} = -\frac{m^2}{4p}$, or at the extremity of the focal ordinate the velocities are changing at the same rate.

2. A point moves in the arc of a circle, its horizontal velocity being 9. Find the accelerations in the path and along Y at the point $x = 3$, the radius of the circle being 5. From

$$x^2 + y^2 = R^2, \quad \frac{dy}{dt} = -\frac{x}{y}\frac{dx}{dt} = -\frac{9x}{y} = -\frac{9x}{\sqrt{R^2 - x^2}},$$

since by condition $\dfrac{dx}{dt} = 9$. Differentiating and dividing by dt,

$$\frac{d^2x}{dt^2} = 0, \quad \frac{d^2y}{dt^2} = -\frac{81\,R^2}{y^3} = -\frac{81\,R^2}{(R^2 - x^2)^{\frac{3}{2}}}, \qquad (1)$$

or the acceleration along X is zero, as it should be since the motion in this direction is uniform, and that along Y is decreasing or increasing algebraically as y is positive or negative.

To find the acceleration in the path, we have

$$\frac{ds}{dt} = \sqrt{\left(\frac{dx}{dt}\right)^2 + \left(\frac{dy}{dt}\right)^2} = \sqrt{1 + \frac{x^2}{y^2}}\,\frac{dx}{dt} = \frac{9\,R}{y},$$

whence $\qquad \dfrac{d^2s}{dt^2} = -\dfrac{9\,R}{y^2}\dfrac{dy}{dt} = \dfrac{81\,Rx}{y^3} = \dfrac{81\,Rx}{(R^2 - x^2)^{\frac{3}{2}}}. \qquad (2)$

Making $x = 3$, $R = 5$, in (1) and (2),

$$\frac{d^2y}{dt^2} = -31.+ \quad \frac{d^2s}{dt^2} = 19.+$$

3. A point moves in the arc of a parabola, the velocity in the direction of Y being constant. Find $\dfrac{dx}{dt}$, and $\dfrac{d^2s}{dt^2}$.

$$\frac{dy}{dt} = m, \quad \frac{dx}{dt} = \frac{my}{p}, \quad \frac{ds}{dt} = \frac{m}{p}\sqrt{p^2 + y^2}, \quad \frac{d^2s}{dt^2} = \frac{m^2y}{p\sqrt{p^2 + y^2}}.$$

The Development of Continuous Functions.

61. Limit of a variable. *The limit of a variable is that value which it constantly approaches but never reaches.*

Thus, the limit of $x = 1 + \frac{1}{2} + \frac{1}{4} + \frac{1}{8} + \cdots$ is 2.

The statement that 2 is the limit of x implies a particular *law of increase*. If x increases by the successive additions of $\frac{1}{4}$ to 1, 2 is not the limit of $x = 1 + \frac{1}{4} + \frac{1}{4} + \cdots$, for by the law of its increase x can be made to exceed 2 in value. But $x = 1 + \frac{1}{2} + \frac{1}{4} + \frac{1}{8} + \cdots$ can never become equal to 2, since by the law of its change each increment is but half the difference between 2 and the value of x at any instant. So if a circle be circumscribed about a regular polygon, its area is not the limit of the area of the polygon if the polygon changes by the motion of its vertices along the produced radii ; for in that case the area of the polygon may become greater than that of the circle. But if the number of sides of an inscribed polygon be indefinitely increased, its vertices remaining in the circle, the area of the circle is the limit of that of the polygon, since no inscribed polygon, however many its sides, can coincide with the circle.

It is evident that if we conceive the law of change of a variable to continue indefinitely in operation, the variable may be made to approach as nearly as we please to its limit. Hence *the difference between a variable and its limit is itself a variable whose limit is zero.*

62. The term limit is also applied to a magnitude of varying position as well as to one of varying value. Thus, OT, the tangent to MN at O, is said to be the limit of the secant OP, since the secant, having at least two points in common with the curve by definition, can never coincide with the tangent; or, more properly, θ is the limit of ϕ as P approaches O. Observe that, as in the previous illustrations, if P approaches O without condition, θ is not the limit of ϕ; but if we affix the condition 'OP remaining a secant,' then θ is the limit of ϕ, P being made to approach as near as we please to O but not coinciding with it.

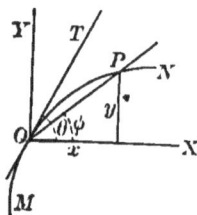

Fig. 10.

63. The term limit is frequently used with another meaning which must be carefully distinguished from that above explained. Thus $\pm R$ are said to be the limiting values of x and y in the equation $x^2 + y^2 = R^2$. To distinguish such limiting values of a variable from one which the variable approaches but never reaches, the latter is often written $x \doteq 2$, $\phi \doteq \theta$, which in the illustrations of Arts. 61 and 62 are read 'x approaches 2 as a limit,' as the number of terms of the series increases indefinitely, 'ϕ approaches θ as a limit,' as P approaches O (Fig. 10).

64. It is evident from the definition that a quantity cannot approach two limits simultaneously. Thus, if 2 is the limit of $x = 1 + \frac{1}{2} + \frac{1}{4} + \cdots$, x can be made to approach 2 in value as near as we please, and therefore no value less than 2 can be its limit; nor can any value greater than 2 be its limit, since it can never equal 2 and therefore cannot be made to approach any value greater than 2 as near as we please.

65. Continuous functions. A function of a variable is **continuous** between certain values of the variable when it has a finite value for every intermediate value of the variable and changes gradually as the variable so changes from one value to the other.

Thus, in $y = mx + b$, y is a continuous function of x for all values of x; in $a^2y^2 + b^2x^2 = a^2b^2$, y is continuous between $x = \pm a$; in $a^2y^2 - b^2x^2 = -a^2b^2$, y is discontinuous between $x = \pm a$, and continuous for values of $x > a$ numerically; in $xy = m$, y is discontinuous for $x = 0$. And, in general, if y is a continuous function for all values of x, $y = f(x)$ represents a curve of unbroken extent.

66. Series. *A succession of terms which follow each other according to some law is called a series.* When known, the law enables us to determine any term of the series.

A series is **finite** or **infinite** as the number of its terms is *limited* or *unlimited.*

67. *The* **sum** *of a* **finite** *series is the sum of its terms.*

The **sum** *of an* **infinite** *series is that finite limit whose value the sum of its terms continually approaches as the number of terms increases.* If there be no such finite value, the series is **divergent**; if such a value exists, the series is **convergent.**

68. *To* **develop** *a function is to find a series whose sum is equal to the function.* The development of a function is therefore a finite or an infinite converging series; in the former case the function being the sum of the terms, and in the latter the limit of the sum of the terms.

When the series is converging, the difference between the function and the sum of the first $n + 1$ terms of the series is called the **remainder after $n + 1$ terms**, and the limit of this remainder as n increases must evidently be zero.

ILLUSTRATIONS. A function may be developed by *involution* when its exponent is a positive integer. Thus $(1 + x)^3 = 1 + 3x + 3x^2 + x^3$, a finite series, whose sum is equal to the function, and which is therefore its development.

A function may be developed by *division* if the indicated division can be completed. Thus, $\dfrac{x^3 - 1}{x - 1} = x^2 + x + 1$, a finite series. When the divisor is not exactly contained in the dividend, division leads to an infinite series, as $\dfrac{1}{1 - x} = 1 + x + x^2 + x^3 + \cdots$, and the process also furnishes the remainder after $n + 1$ terms. Since this remainder, when added to the terms already found, must equal the function, it must decrease as n increases, and its examination will discover whether the series is or is not converging, that is, whether it is or is not the development of the function. Thus, in the above case, the remainder after $n + 1$ terms is $\dfrac{x^{n+1}}{1 - x}$, which decreases as n increases, only when $x < 1$. Hence if $x < 1$, the series is converging, and we may write $\dfrac{1}{1 - x} = 1 + x + x^2 + x^3 + \cdots$, understanding that the second number approximates more closely in value to the first as the series is extended; while if $x > 1$, the series is diverging, and cannot be equal to the function, or is not its development.

Other processes of deriving a series from a function do not afford the remainder, and thus do not indicate whether the series diverges or converges. Thus *evolution*, or the extraction of the root of a polynomial, furnishes in general an infinite series, but no remainder.

No universal criterion for determining whether a given series is converging or diverging has been found.

69. Maclaurin's theorem. *The object of Maclaurin's theorem is the development of a function of a* **single** *variable into a series arranged according to the ascending powers of the variable with finite and constant coefficients.*

The proposed development will be of the form

$$f(x) = A + Bx + Cx^2 + Dx^3 + Ex^4 + \cdots, \qquad (1)$$

in which A, B, C, etc., are finite and independent of x. It is required to find such values for A, B, C, etc., as will satisfy (1) for all values of x, that is, render the series either finite, or, if infinite, then converging.

Since (1) is to be true for all values of x, it must be true for $x = 0$; whence $A = f(x)$ when $x = 0$, or the first term of the series is what the function becomes when $x = 0$. Differentiating (1), the successive derivatives are

$$f'(x) = B + 2Cx + 3Dx^2 + 4Ex^3 + \cdots,$$

$$f''(x) = 2C + 2 \cdot 3Dx + 3 \cdot 4Ex^2 + \cdots,$$

$$f'''(x) = 2 \cdot 3D + 2 \cdot 3 \cdot 4Ex + \cdots,$$

$$\text{etc.,}$$

which, being true for all values of x, are true for $x = 0$. Hence representing by $f(0), f'(0), f''(0)$, etc., what $f(x), f'(x), f''(x)$, etc., become when $x = 0$, we have

$$A = f(0),$$

$$B = f'(0),$$

$$2C = f''(0), \quad \therefore \ C = \frac{f''(0)}{\underline{|2}},$$

$$2 \cdot 3D = f'''(0), \quad \therefore \ D = \frac{f'''(0)}{\underline{|3}},$$

$$\text{etc.,} \qquad\qquad \text{etc.,}$$

and substituting these values in (1),

$$f(x) = f(0) + f'(0)x + f''(0)\frac{x^2}{\underline{|2}} + f'''(0)\frac{x^3}{\underline{|3}} + \cdots, \qquad (2)$$

and the theorem may be thus stated:

The first term of the series is what the function becomes when $x = 0$; the second term is what the first derivative of the function becomes when $x = 0$, into x; the third term is what the second derivative of the function becomes when $x = 0$, into x^2 divided by factorial 2; and, in general, the $(n+1)$th term is what the nth derivative of the function becomes when $x = 0$, into x^n divided by factorial n.

If the resulting series is finite, it is equal to the function, the two members of (2) are identical, and the development is effected.   If the resulting series is infinite, it is necessary to determine whether it is convergent.

70. Taylor's theorem.   *The object of Taylor's theorem is the development of a function of the* **algebraic sum** *of* **two variables** *into a series arranged according to the ascending powers of one of the variables, with finite coefficients depending upon the other and the constants which enter the function.*

The proposed development will be of the form

$$f(x+y) = P + Qy + Ry^2 + Sy^3 + \cdots, \tag{1}$$

in which P, Q, R, etc., are functions of x, and independent of y. It is required to find such values of P, Q, R, etc., as will satisfy (1) for all values of x and y, that is, render the series finite, or, if infinite, then converging.

Since (1) is to be true for all values of x and y, it must be true when $y = 0$; in which case $P = f(x)$, or the first term of the series is what the function becomes when $y = 0$.

Let a be any value of x, and P', Q', R', etc., the corresponding values of the coefficients, which are functions of x.   Then (1) is true for $x = a$, and we have

$$f(a+y) = P' + Q'y + R'y^2 + S'y^3 + \cdots, \tag{2}$$

whose successive derivatives are

$$f'(a+y) = Q' + 2R'y + 3S'y^2 \cdots,$$
$$f''(a+y) = 2R' + 2 \cdot 3 S'y \cdots,$$
$$f'''(a+y) = 2 \cdot 3 S' \cdots,$$

etc.

Since these equations must be true for all values of y, they are true for $y = 0$.   Hence

$$P' = f(a), \quad Q' = f'(a), \quad R' = \frac{f''(a)}{\lfloor 2}, \quad S' = \frac{f'''(a)}{\lfloor 3}, \quad \text{etc.}$$

Substituting these values of P', Q', R', etc., in (2),

$$f(a+y)=f(a)+f'(a)y+f''(a)\frac{y^2}{\underline{2}}+f'''(a)\frac{y^3}{\underline{3}}\cdots,$$

in which the coefficients are what $f(x)$, $f'(x)$, $f''(x)$, etc., become when $x = a$. But a is any arbitrary value; hence, whatever the value of x,

$$f(x+y)=f(x)+f'(x)y+f''(x)\frac{y^2}{\underline{2}}+f'''(x)\frac{y^3}{\underline{3}}\cdots,$$

and the theorem may be thus stated:

The first term of the series is what the function becomes when $y = 0$; the second term is the first derivative of the function when $y = 0$, into y; the third term is the second derivative of the function when $y = 0$, into y^2 divided by factorial 2; and, in general, the $(n+1)$th term is the nth derivative of the function when $y = 0$, into y^n divided by factorial n.

As before, if the series thus obtained is infinite, it is necessary to determine whether it is convergent.

71. Completion of Taylor's and Maclaurin's Formulæ.

Since the use of infinite series as the equivalents of the functions is inadmissible unless the series are converging, it is necessary to determine the remainder after $n+1$ terms in the preceding formulæ, and to examine this remainder in any particular case to see if its limit is zero as n increases.

I. *If $f(x)$ becomes zero when $x = a$ and $x = b$, and is continuous between these values, and if $f'(x)$ is also continuous between these values, then $f'(x)$ will be zero for some value of x between a and b.*

For, since $f(x) = 0$ for $x = a$ and $x = b$, as x changes from a to b, $f(x)$ must either first increase and then decrease, or first decrease and then increase. But the first derivative is positive when the function is increasing and negative when it is decreasing (Art. 22), and therefore in either case it changes sign between the values $x = a$ and $x = b$; and being continuous, it cannot become infinite, and therefore must pass through zero.

II. *First form of the remainder.*

Resuming Taylor's formula,

$$(1)\ f(x+y)=f(x)+f'(x)y+f''(x)\frac{y^2}{\underline{2}}+f''(x)\frac{y^3}{\underline{3}}\cdots+f^n(x)\frac{y^n}{\underline{n}}+\cdots.$$

Writing $x+y = X$, whence $y = X - x$, and representing by R the remainder after $n+1$ terms, we have

$$(2)\ f(X)=f(x)+f'(x)(X-x)+f'(x)\frac{(X-x)^2}{\underline{2}}+f''(x)\frac{(X-x)^3}{\underline{3}}\cdots$$
$$+f^n(x)\frac{(X-x)^n}{\underline{n}}+R.$$

Writing the remainder in the form $P \cdot \dfrac{(X-x)^{n+1}}{\lfloor n+1}$, P being a function of X and x to be determined, substituting this value of R, and transposing, we obtain

(3) $f(X) - f(x) - f'(x)(X-x) - f''(x)\dfrac{(X-x)^2}{\lfloor 2} - f'''(x)\dfrac{(X-x)^3}{\lfloor 3} \cdots$

$$- f^n(x)\dfrac{(X-x)^n}{\lfloor n} - P \cdot \dfrac{(X-x)^{n+1}}{\lfloor n+1} = 0.$$

Representing by $F(z)$ the function of z which (3) becomes by substituting z for x,

(4) $F(z) = f(X) - f(z) - f'(z)(X-z) - f''(z)\dfrac{(X-z)^2}{\lfloor 2} - f'''(z)\dfrac{(X-z)^3}{\lfloor 3} \cdots$

$$- f^n(z)\dfrac{(X-z)^n}{\lfloor n} - P \cdot \dfrac{(X-z)^{n+1}}{\lfloor n+1}.$$

If $z = x$ in (4), it becomes identical with (3) and therefore $= 0$. It also becomes zero if $z = X$, for every term then contains a zero factor. Therefore, by I., its derivative $F'(z)$ must be zero for some value of z between x and X. If θ be a proper fraction, $z = x + \theta(X-x)$ will represent such intermediate value.

Differentiating (4) to obtain $F'(z)$, we have

$F'(z) = 0 - f'(z) + f'(z) - f''(z)(X-z) + f''(z)(X-z)$

$$- f'''(z)\dfrac{(X-z)^2}{\lfloor 2} + f'''(z)\dfrac{(X-z)^2}{\lfloor 2} - \cdots + f^n(z)\dfrac{(X-z)^{n-1}}{\lfloor n-1}$$

$$- f^{n+1}(z)\dfrac{(X-z)^n}{\lfloor n} + P\dfrac{(X-z)^n}{\lfloor n},$$

whose terms vanish in pairs, except the last two, giving

$$F'(z) = - f^{n+1}(z)\dfrac{(X-z)^n}{\lfloor n} + P\dfrac{(X-z)^n}{\lfloor n}.$$

Substituting the value $z = x + \theta(X-x)$ for which $F'(z)$ is zero, we have, after cancelling the common factor $\dfrac{(X-z)^n}{\lfloor n}$,

or $$- f^{n+1}[x + \theta(X-x)] + P = 0,$$

$$P = f^{n+1}[x + \theta(X-x)],$$

in which all we know of θ is that its value lies between 0 and 1.

Hence the remainder after $n+1$ terms is

$$R = P\dfrac{(X-x)^{n+1}}{\lfloor n+1} = f^{n+1}[x + \theta(X-x)]\dfrac{(X-x)^{n+1}}{\lfloor n+1} = f^{n+1}(x+\theta y)\dfrac{y^{n+1}}{\lfloor n+1},$$

and, substituting in (1), the completed form of Taylor's theorem is

(5) $f(x+y) = f(x) + f'(x)y + f''(x)\dfrac{y^2}{\lfloor 2} \cdots + f^n(x)\dfrac{y^n}{\lfloor n} + f^{n+1}(x+\theta y)\dfrac{y^{n+1}}{\lfloor n+1}.$

Making $x = 0$, and changing y to x,

(6) $f(x) = f(0) + f'(0)x + f''(0)\dfrac{x^2}{\lfloor 2} \cdots + f^n(0)\dfrac{x^n}{\lfloor n} + f^{n+1}(\theta x)\dfrac{x^{n+1}}{\lfloor n+1},$

the completed form of Maclaurin's theorem.

We thus have in both cases the remainder after $n+1$ terms, which is found by differentiating the function $n + 1$ times and changing x to $x + \theta y$, or to θx, in the $(n + 1)$th derivative, and multiplying this result by $\dfrac{y^{n+1}}{\underline{n+1}}$, or by $\dfrac{x^{n+1}}{\underline{n+1}}$. *If this remainder is zero, the series is finite; if its limit is zero as n increases, the series is convergent; if it is neither zero nor has zero for a limit, the formulæ fail.*

III. *Second form of the remainder.*

Writing the remainder in the form $R = P_1(X - x)$, (3) would become

(7) $f(X) - f(x) - f'(x)(X - x) - f'(x)\dfrac{(X - x)^2}{\underline{2}} \cdots - f^n(x)\dfrac{(X - x)^n}{\underline{n}} - P_1(X - x) = 0.$

Representing by $F(z)$ what (7) becomes by substituting z for x,

$F(z) = f(X) - f(z) - f'(z)(X - z) - f''(z)\dfrac{(X - z)^2}{\underline{2}} \cdots - f^n(z)\dfrac{(X - z)^n}{\underline{n}} - P_1(X - z),$

in which, if $z = x$ or $z = X$, $F(z) = 0$ as before; and therefore $F(z) = 0$ for $z = x + \theta(X - x)$. Differentiating to find $F'(z)$, the terms vanish in pairs except the last two, giving

$$F'(z) = -f^{n+1}(z)\frac{(X - z)^n}{\underline{n}} + P_1,$$

and, substituting the value of z for which $F'(z) = 0$,

$$-f^{n+1}[x + \theta(X - x)]\frac{[X - (x + \theta(X - x))]^n}{\underline{n}} + P_1 = 0,$$

or

$$P_1 = f^{n+1}[x + \theta(X - x)]\frac{[X - x]^n(1 - \theta)^n}{\underline{n}} = f^{n+1}(x + \theta y)\frac{y^n(1 - \theta)^n}{\underline{n}},$$

and

$$R = P_1(X - x) = f^{n+1}(x + \theta y)\frac{y^{n+1}(1 - \theta)^n}{\underline{n}}.$$

Substituting in (1), a second completed form of Taylor's formula is

(8) $f(x + y) = f(x) + f'(x)y + f'(x)\dfrac{y^2}{\underline{2}} \cdots + f^n(x)\dfrac{y^n}{\underline{n}} + f^{n+1}(x + \theta y)\dfrac{y^{n+1}(1 - \theta)^n}{\underline{n}}.$

Making $x = 0$ and changing y to x, the corresponding form of Maclaurin's formula is

(9) $f(x) = f(0) + f'(0)x + f'(0)\dfrac{x^2}{\underline{2}} \cdots + f^n(0)\dfrac{x^n}{\underline{n}} + f^{n+1}(\theta x)\dfrac{x^{n+1}(1 - \theta)^n}{\underline{n}},$

to which forms apply the remarks made upon (5) and (6).

IV. *If the $(n + 1)$th derivative is finite for all values of n, Taylor's and Maclaurin's formulæ develop $f(x + y)$ and $f(x)$, respectively.*

The first forms of the remainder are

$$R = f^{n+1}(x + \theta y)\frac{y^{n+1}}{\underline{n+1}} \text{ and } R = f^{n+1}(\theta x)\frac{x^{n+1}}{\underline{n+1}}.$$

But when $n + 1$, as n increases, becomes equal to x, $\dfrac{x^{n+1}}{\underline{n+1}}$ begins and continues to diminish, each successive value being less than the preceding one. Hence, whatever the value of x, provided only it be finite, as it is by hypothesis, $\dfrac{x^{n+1}}{\underline{n+1}}$ tends to the limit zero as n increases indefinitely. It follows, therefore, that if the $(n + 1)$th derivative does not become infinite with n, R approaches zero as n increases, and the series is convergent.

The same is also true of the second forms of R.

V. It is evident that the sum of the first n terms of a series cannot approach a fixed value as n increases indefinitely, unless the terms finally decrease; that is, *unless the ratio of the nth term to the one before it becomes and continues less than unity as n increases, the series cannot be convergent.*

72. Applications. Assuming that the following functions can be developed, show that:

1. $(a + x)^7 = a^7 + 7\,a^6x + 21\,a^5x^2 + 35\,a^4x^3 + 35\,a^3x^4 + 21\,a^2x^5$
$$+ 7\,ax^6 + x^7.$$

Making $x = 0$, $f(0) = a^7$.

The successive derivatives are:

$f'(x) = 7(a + x)^6$,   whence $f'(0) = 7a^6$.

$f''(x) = 6 \cdot 7(a + x)^5$,   " $f''(0) = 6 \cdot 7a^5$.

$f'''(x) = 5 \cdot 6 \cdot 7(a + x)^4$,   " $f'''(0) = 5 \cdot 6 \cdot 7a^4$.

$f^{iv}(x) = 4 \cdot 5 \cdot 6 \cdot 7(a + x)^3$,   " $f^{iv}(0) = 4 \cdot 5 \cdot 6 \cdot 7a^3$.

$f^{v}(x) = 3 \cdot 4 \cdot 5 \cdot 6 \cdot 7(a + x)^2$,   " $f^{v}(0) = 3 \cdot 4 \cdot 5 \cdot 6 \cdot 7a^2$.

$f^{vi}(x) = 2 \cdot 3 \cdot 4 \cdot 5 \cdot 6 \cdot 7(a + x)$, " $f^{vi}(0) = 2 \cdot 3 \cdot 4 \cdot 5 \cdot 6 \cdot 7a$.

$f^{vii}(x) = 2 \cdot 3 \cdot 4 \cdot 5 \cdot 6 \cdot 7$,   " $f^{vii}(0) = 2 \cdot 3 \cdot 4 \cdot 5 \cdot 6 \cdot 7$.

$f^{viii}(x) = 0$.

Substituting in Maclaurin's formula,

$$f(x) = f(0) + f'(0)x + f''(0)\frac{x^2}{\underline{2}} + f'''(0)\frac{x^3}{\underline{3}} \cdots,$$

we have

$$(a + x)^7 = a^7 + 7\,a^6x + 6 \cdot 7a^5\frac{x^2}{\underline{2}} + 5 \cdot 6 \cdot 7a^4\frac{x^3}{\underline{3}} + 4 \cdot 5 \cdot 6 \cdot 7a^3\frac{x^4}{\underline{4}}$$

$$+ 3 \cdot 4 \cdot 5 \cdot 6 \cdot 7a^2\frac{x^5}{\underline{5}} + 2 \cdot 3 \cdot 4 \cdot 5 \cdot 6 \cdot 7a\frac{x^6}{\underline{6}}$$

$$+ 2 \cdot 3 \cdot 4 \cdot 5 \cdot 6 \cdot 7\frac{x^7}{\underline{7}}$$

$$= a^7 + 7\,a^6x + 21\,a^5x^2 + 35\,a^4x^3 + 35\,a^3x^4 + 21\,a^2x^5$$
$$+ 7\,ax^6 + x^7.$$

λ

Being finite, the series is the development of the function, as will evidently be the case so long as the exponent of the binomial is a positive integer.

2. $\sin x = x - \dfrac{x^3}{\underline{|3}} + \dfrac{x^5}{\underline{|5}} - \dfrac{x^7}{\underline{|7}} \cdots$

Making $x = 0$, $f(0) = 0$.

The successive derivatives are.

$$f'(x) = \cos x, \quad \text{whence} \quad f'(0) = 1.$$
$$f''(x) = -\sin x, \quad \text{``} \quad f''(0) = 0.$$
$$f'''(x) = -\cos x, \quad \text{``} \quad f'''(0) = -1.$$
$$f^{iv}(x) = \sin x, \quad \text{``} \quad f^{iv}(0) = 0.$$

Since $f^{iv}(x)$ is the original function, these values will recur in sets of four, and we have

$$\sin x = x - \dfrac{x^3}{\underline{|3}} + \dfrac{x^5}{\underline{|5}} - \dfrac{x^7}{\underline{|7}} \cdots.$$

3. $\cos x = 1 - \dfrac{x^2}{\underline{|2}} + \dfrac{x^4}{\underline{|4}} - \dfrac{x^6}{\underline{|6}} + \cdots.$

Since the $(n + 1)$th derivatives of $\sin x$ and $\cos x$ are finite whatever the value of n, the formula develops these functions (Art. 71, IV.), and the error may be made as small as we please by taking a sufficient number of terms.

By means of these series we may compute the natural sine or cosine of any arc, but few terms being necessary as the series converge rapidly. Thus, if $x = \dfrac{\pi}{18} = .174533$ be substituted for x in the series of Ex. 2, $\sin x = \sin 10° = .17365+$.

4. $a^x = 1 + \log a \cdot x + \log^2 a \dfrac{x^2}{\underline{|2}} + \log^3 a \dfrac{x^3}{\underline{|3}} + \cdots.$

Making $x = 0$, $f(0) = a^0 = 1$.

The successive derivatives are

$$f'(x) = a^x \log a, \quad f''(x) = a^x \log^2 a, \quad f'''(x) = a^x \log^3 a, \quad \text{etc.} ;$$

whence

$$f'(0) = \log a, \quad f''(0) = \log^2 a, \quad f'''(0) = \log^3 a, \quad \text{etc.,}$$

which, substituted in Maclaurin's formula, give the above series. Making $a = e$, whence $\log e = 1$, it becomes

$$e^x = 1 + x + \frac{x^2}{\underline{|2}} + \frac{x^3}{\underline{|3}} + \frac{x^4}{\underline{|4}} \cdots,$$

and if $x = 1$,

$$e = 1 + 1 + \frac{1}{\underline{|2}} + \frac{1}{\underline{|3}} + \frac{1}{\underline{|4}} \cdots = 2.718281+$$

the *Naperian base.* These are the *exponential series.*

The $(n + 1)$th derivative of a^x is $(\log a)^{n+1} a^x$, and hence

$$R = f^{n+1}(\theta x) \frac{x^{n+1}}{\underline{|n+1}} = \frac{(x \log a)^{n+1}}{\underline{|n+1}} a^{\theta x}.$$

But $a^{\theta x}$ is finite, and

$$\frac{(x \log a)^{n+1}}{\underline{|n+1}} = \frac{x \log a}{1} \cdot \frac{x \log a}{2} \cdots \frac{x \log a}{n+1},$$

which approaches zero as n increases; therefore the formula develops a^x.

5. $(1 + x)^m = 1 + mx + \dfrac{m(m-1)}{\underline{|2}} x^2 + \dfrac{m(m-1)(m-2)}{\underline{|3}} x^3 \cdots$

$$+ \frac{m(m-1) \cdots (m-n+1)}{\underline{|n}} x^n + \cdots.$$

To determine for what values of x the formula develops $(1 + x)^m$.

The $(n + 1)$th derivative, when θx is written for x, is

$$m(m-1) \cdots (m-n)(1 + \theta x)^{m-n-1},$$

which becomes zero if m is a positive integer when $n = m$. Hence the series is finite, and is the development of $(1 + x)^m$ when m is a positive integer. If m is negative or fractional, the series is infinite. The ratio of its nth term to the one immediately before it is

$$\frac{m-n+1}{n} x = \left(\frac{m+1}{n} - 1 \right) x,$$

whose absolute value, as n increases, will eventually become and remain greater than unity if x is numerically greater than 1. Hence (Art. 71, V.) the series is divergent, and cannot equal $(1 + x)^m$ when x is numerically greater than 1. The remainder after $n + 1$ terms is

$$R = f^{n+1}(\theta x) \frac{x^{n+1}}{\underline{|n+1}} = \left[\frac{m(m-1) \cdots (m-n)}{\underline{|n+1}} x^{n+1} \right] \frac{1}{(1 + \theta x)^{n-m+1}}.$$

When x lies between 0 and 1, the last factor becomes less than 1 as n increases. Increasing n by 1 multiplies the first factor by $\dfrac{m-n-1}{n+2} x$, or $\left(\dfrac{m-1}{n+2} - \dfrac{n}{n+2} \right) x$, which approaches $-x$ as n increases; that is, a quantity numerically less than 1. Hence to increase n indefinitely is to multiply by an infinite number of factors each less than 1; the product therefore decreases indefinitely, and the formula develops $(1 + x)^m$ for values of x between 0 and 1. By means of the second form of the remainder we have

$$R = f^{n+1}(\theta x)\, \frac{x^{n+1}(1-\theta)^n}{\lfloor n} = \left[\frac{m(m-1)\cdots(m-n)}{\lfloor n} x^{n+1}\right](1+\theta x)^{m-n-1}(1-\theta)^n$$

$$= \left[\frac{m(m-1)\cdots(m-n)}{\lfloor n} x^{n+1}\right]\left(\frac{1-\theta}{1+\theta x}\right)^{n+1}\frac{(1+\theta x)^m}{1-\theta}.$$

When x lies between 0 and -1, the last factor is finite; $\left(\dfrac{1-\theta}{1+\theta x}\right)^{n+1}$ approaches zero as n increases; increasing n by 1 multiplies the first factor by $\dfrac{m-n-1}{n+1}x$, which approaches $-x$ as n increases. Hence, as before, the formula develops $(1+x)^m$ for values of x between 0 and -1.

Since $(a+x)^m$ may be written in either of the forms

$$a^m\left(1+\frac{x}{a}\right)^m, \quad x^m\left(1+\frac{a}{x}\right)^m,$$

and as *one* of those can be developed, whatever the relative values of a and x, the Binomial formula holds good for fractional and negative exponents. When m is a positive integer, the series is finite, and the formula holds good for *both* the above forms.

6. $\log(1+x) = x - \dfrac{x^2}{2} + \dfrac{x^3}{3} - \dfrac{x^4}{4}\cdots$.

Making $x = 0, f(0) = \log 1 = 0$.

The successive derivatives are

$$f'(x) = \frac{1}{1+x}, \qquad f''(x) = -\frac{1}{(1+x)^2},$$

$$f'''(x) = \frac{2}{(1+x)^3}, \qquad f^{iv}(x) = -\frac{2\cdot 3}{(1+x)^4}, \text{ etc. };$$

whence $f'(0) = 1, f''(0) = -1, f'''(0) = 2, f^{iv}(0) = -2\cdot 3$, etc., and these in Maclaurin's formula give the above series.

The ratio of the nth term to the preceding one is $\dfrac{(-1)(n-1)}{n}x$ or $-\left(1-\dfrac{1}{n}\right)x$, which, if x is numerically greater than 1, becomes and remains greater than unity as n increases; hence (Art. 71, V.) the series is divergent if x is numerically greater than 1. The $(n+1)$th derivative is $(-1)^n \dfrac{\lfloor n}{(1+x)^{n+1}}$, and, using the first form of R,

$$R = f^{n+1}(\theta x)\frac{x^{n+1}}{\lfloor n+1} = \frac{(-1)^n}{n+1}\left(\frac{x}{1+\theta x}\right)^{n+1}.$$

If x lies between 0 and $+1$, $\dfrac{x}{1+\theta x}$ is a proper fraction, and R approaches zero as n increases.

If x lies between 0 and -1, the series becomes $-x - \dfrac{x^2}{2} - \dfrac{x^3}{3}\cdots$, and the second form of R gives, numerically,

$$R = f^{n+1}(\theta x)\frac{x^{n+1}(1-\theta)^n}{\lfloor n} = \left(\frac{x-\theta x}{1-\theta x}\right)^n\frac{x}{1-\theta x}.$$

For values of x between 0 and 1, $\left(\frac{x - \theta x}{1 - \theta x}\right)^n$ is a proper fraction, and approaches zero as n increases, while the last factor is finite. Hence the formula develops $\log(1+x)$ when x lies between $+1$ and -1.

7. $\log_a(1+x) = m\left(x - \frac{x^2}{2} + \frac{x^3}{3} - \frac{x^4}{4} + \frac{x^5}{5} \cdots\right);$ (1)

if $a = e$, we have, as in Ex. 6,

$\log(1+x) = x - \frac{x^2}{2} + \frac{x^3}{3} - \frac{x^4}{4} + \frac{x^5}{5} \cdots,$ (2)

which are the *logarithmic series*. As they diverge, if $x > 1$, they are not suitable for the computation of logarithms. To adapt them to this purpose, substitute $-x$ for x in (1), and we have

$\log_a(1-x) = m\left(-x - \frac{x^2}{2} - \frac{x^3}{3} - \frac{x^4}{4} - \frac{x^5}{5} \cdots\right).$ (3)

Subtracting (3) from (1),

$\log_a(1+x) - \log_a(1-x) = 2m\left\{x + \frac{x^3}{3} + \frac{x^5}{5} + \frac{x^7}{7} + \cdots\right\}.$

Let $x = \dfrac{1}{2z+1}$; then x is less than 1 for all positive values of z, and

$$\log_a(1+x) - \log_a(1-x) = \log_a\frac{1+x}{1-x} = \log_a\frac{z+1}{z}$$
$$= \log_a(z+1) - \log_a z$$
$$= 2m\left\{\frac{1}{2z+1} + \frac{1}{3(2z+1)^3} + \frac{1}{5(2z+1)^5} \cdots\right\}, \quad (4)$$

or, if $a = e$, whence $m = 1$,

$$\log(z+1) - \log z = 2\left\{\frac{1}{2z+1} + \frac{1}{3(2z+1)^3} + \frac{1}{5(2z+1)^5} \cdots\right\}.$$

From this series, which converges rapidly, we may compute the Naperian logarithms of numbers. Thus, if $z = 1$, $\log 1 = 0$, and we have

$$\log 2 = 2\left\{\frac{1}{3} + \frac{1}{3\cdot 3^3} + \frac{1}{5\cdot 3^5} + \frac{1}{7\cdot 3^7} + \cdots\right\} = .693147+,$$

when six terms are taken.

Making $z = 2$,

$$\log 3 = \log 2 + 2 \left\{ \frac{1}{5} + \frac{1}{3 \cdot 5^3} + \frac{1}{5 \cdot 5^5} + \frac{1}{7 \cdot 5^7} + \cdots \right\} = 1.098612+.$$

$$\log 4 = 2 \log 2 = 1.386294+.$$

Making $x = 4$,

$$\log 5 = \log 4 + 2 \left\{ \frac{1}{9} + \frac{1}{3 \cdot 9^3} + \frac{1}{5 \cdot 9^5} + \frac{1}{7 \cdot 9^7} + \cdots \right\} = 1.6094379+.$$

In like manner, the Naperian logarithms of all numbers may be computed.

Cor. 1. *The Naperian logarithm of the base of the common system is*
$$\log 10 = \log 5 + \log 2 = 2.302585+.$$

Cor. 2. From (4), b being the base of the system, and m' the corresponding modulus,

$$\log_b \frac{z+1}{z} = 2m' \left\{ \frac{1}{2z+1} + \frac{1}{3(2z+1)^3} + \cdots \right\}. \quad (5)$$

Since (4) and (5) are true for all positive values of z, writing x for $\frac{z+1}{z}$, we have

$$\frac{\log_a x}{\log_b x} = \frac{m}{m'}, \quad (6)$$

or *the logarithms of the same number in different systems are proportional to the moduli of the systems.*

Cor. 3. If in (6) $b = e$, then $m' = 1$, and

$$\log_a x = m \log x. \quad (7)$$

Having then computed, as above, a table of Naperian logarithms, the *logarithms in any system may be found by multiplying their Naperian logarithms by the modulus of the system.*

Cor. 4. Since $\log_a a = 1$, if $x = a$ in (7),

$$m = \frac{1}{\log a},$$

or *the modulus of any system is the reciprocal of the Naperian logarithm of its base;* which is the relation between the modulus of a system and its base referred to in Art. 28.

Cor. 5. In the common system $a = 10$, hence

$$m = \frac{1}{\log 10} = \frac{1}{2.302585} = .434294+,$$

the *modulus of the common system.*

8. $\tan^{-1} x = x - \dfrac{x^3}{3} + \dfrac{x^5}{5} - \dfrac{x^7}{7} \cdots$.

Making $x = 0$, $f(0) = 0$.

The first derivative is $\dfrac{1}{1+x^2} = 1 - x^2 + x^4 - x^6 + x^8 - x^{10} \cdots$ by division; hence the successive derivatives are

$$f'(x) = 1 - x^2 + x^4 - x^6 + x^8 - x^{10} \cdots,$$
$$f''(x) = -2x + 4x^3 - 6x^5 + 8x^7 - 10x^9 \cdots,$$
$$f'''(x) = -2 + 3\cdot4x^2 - 5\cdot6x^4 + 7\cdot8x^6 - 9\cdot10x^8 \cdots,$$
$$f^{iv}(x) = 2\cdot3\cdot4x - 4\cdot5\cdot6x^3 + 6\cdot7\cdot8x^5 - 8\cdot9\cdot10x^7 \cdots,$$
$$f^{v}(x) = 2\cdot3\cdot4 - 3\cdot4\cdot5\cdot6x^2 + \cdots,$$

from which

$$f'(0) = 1, \qquad f'''(0) = -2, \qquad f^{v}(0) = 2\cdot3\cdot4,$$
$$f''(0) = 0, \qquad f^{iv}(0) = 0, \qquad \text{etc.},$$

and these in Maclaurin's formula give the series above.

Since a series whose terms are alternately plus and minus converges if each term is numerically less than the preceding, the series converges for $x = 1$, whence $\tan^{-1}1 = 45° = \dfrac{\pi}{4}$, and we have

$$\pi = 4(1 - \tfrac{1}{3} + \tfrac{1}{5} - \tfrac{1}{7} + \tfrac{1}{9} \cdots),$$

whence the value of π.

9. $\sin(x + y) = \sin x \cos y + \cos x \sin y$.

This being a function of the sum of two variables, we use Taylor's formula.

Making $y = 0$, $f(x) = \sin x$, whose successive derivatives are

$$f'(x) = \cos x, \quad f''(x) = -\sin x, \quad f'''(x) = -\cos x, \quad f^{iv}(x) = \sin x,$$

and so on in sets of four. Hence, substituting in Taylor's formula,

$$\sin(x+y) = \sin x + \cos x \cdot y - \sin x \frac{y^2}{\underline{|2}} - \cos x \frac{y^3}{\underline{|3}} + \sin x \frac{y^4}{\underline{|4}} + \cos x \frac{y^5}{\underline{|5}} \cdots$$

$$= \sin x \left\{ 1 - \frac{y^2}{\underline{|2}} + \frac{y^4}{\underline{|4}} - \cdots \right\} + \cos x \left\{ y - \frac{y^3}{\underline{|3}} + \frac{y^5}{\underline{|5}} - \cdots \right\}$$

$$= \sin x \cos y + \cos x \sin y \quad \text{(Exs. 2 and 3)}.$$

10. $\cos (x + y) = \cos x \cos y - \sin x \sin y.$

11. $\sin (x - y) = \sin x \cos y - \cos x \sin y.$

12. $\cos (x - y) = \cos x \cos y + \sin x \sin y.$

13. Deduce the Binomial formula by Taylor's theorem from $(x + y)^m$.

Making $y = 0$, $f(x) = x^m$, whose successive derivatives are

$$f'(x) = m x^{m-1}, \quad f''(x) = m(m-1) x^{m-2}, \quad \text{etc.},$$

hence $\quad (x + y)^m = x^m + m x^{m-1} y + m(m-1) x^{m-2} \frac{y^2}{\underline{|2}} + \text{etc.}$

14. $e^{\sin x} = 1 + x + \dfrac{x^2}{\underline{|2}} - \dfrac{3 x^4}{\underline{|4}} + \dfrac{8 x^5}{\underline{|5}} - \dfrac{3 x^6}{\underline{|6}} \cdots.$

15. $e^{\cos x} = e \left\{ 1 - \dfrac{x^2}{2} + \dfrac{4 x^4}{\underline{|4}} - \dfrac{31 x^6}{\underline{|6}} \cdots \right\}.$

16. $\tan x = x + \dfrac{x^3}{3} + \dfrac{2 x^5}{15} \cdots.$

17. $\sec x = 1 + \dfrac{x^2}{2} + \dfrac{5 x^4}{24} \cdots.$

18. $\dfrac{e^x}{\cos x} = 1 + x + x^2 + \dfrac{2 x^3}{3} + \dfrac{x^4}{2} \cdots.$

19. $x^2 e^x = x^2 + x^3 + \dfrac{x^4}{\underline{|2}} + \dfrac{x^5}{\underline{|3}} \cdots.$

20. $e^{x \sin x} = 1 + x^2 + \dfrac{x^4}{3} \cdots$.

21. $e^{\tan^{-1} x} = 1 + x + \dfrac{x^2}{2} - \dfrac{x^3}{6} - \dfrac{7 x^4}{24} \cdots$.

22. $\log_a(x + y) = \log_a x + m \left\{ \dfrac{y}{x} - \dfrac{y^2}{2 x^2} + \dfrac{y^3}{3 x^3} - \dfrac{y^4}{4 x^4} + \cdots \right\}$.

23. $a^{x+y} = a^x \left\{ 1 + \log a \cdot y + \log^2 a \dfrac{y^2}{\lfloor 2} + \log^3 a \dfrac{y^3}{\lfloor 3} \cdots \right\}$.

24. $\sin^{-1}(x + y) = \sin^{-1} x + \dfrac{y}{(1 - x^2)^{\frac{1}{2}}} + \dfrac{x y^2}{\lfloor 2 \, (1 - x^2)^{\frac{3}{2}}}$

$\qquad\qquad + \dfrac{(1 + 2 x^2) y^3}{\lfloor 3 \, (1 - x^2)^{\frac{5}{2}}} \cdots$.

25. $(a^2 - e^2 x^2)^{\frac{1}{2}} = a - \dfrac{e^2 x^2}{2 a} - \dfrac{e^4 x^4}{2 \cdot 4 a^3} - \dfrac{3 e^6 x^6}{2 \cdot 4 \cdot 6 a^5} - \cdots$.

73. *Failing cases of Maclaurin's and Taylor's formulæ.*

It has been seen that the above formulæ often lead to diverging series and therefore fail. The following exceptions are also to be noted.

Since the proof that the formulæ develop any function depends upon the condition that the derivatives of the functions are continuous, no one of them becoming infinite for a finite value of the variable, if $\log x$ be the function, whose first derivative $f'(x) = \dfrac{1}{x}$ becomes ∞, as do all the succeeding derivatives, when $x = 0$, the coefficients $f'(0)$, $f''(0)$, etc., of Maclaurin's formula become infinite, the series has no determinate value, and $\log x$ cannot be developed in powers of x. The same is true of $x^{\frac{1}{2}}$, $a^{\frac{1}{x}}$, cosec x, cot x, etc.

Again, from $(x+y+a)^{\frac{1}{2}}$, we have, for $y = 0$, $f(x) = (x + a)^{\frac{1}{2}}$, whence $f'(x) = \dfrac{1}{2(x + a)^{\frac{1}{2}}}$, which is finite for all values of x except $x = -a$. For this value of x, $f'(x) = \infty$, as are all the successive derivatives. Hence the coefficients $f'(x)$, $f''(x)$, etc., of Taylor's formula become infinite for $x = -a$, and the function $(x + y + a)^{\frac{1}{2}}$ can be developed in powers of y for all values of x except $x = -a$.

Evaluation of Illusory Forms.

74. The form $\dfrac{0}{0}$. It frequently happens that for a particular value of the variable a function assumes the form $\dfrac{0}{0}$. Thus $\dfrac{\sin x}{x} = \dfrac{0}{0}$ when $x = 0$. How is this result to be interpreted?

Let x, y, be the coordinates of P, x and y being functions of z, and let MN be the curve the coordinates of whose points are the simultaneous values of x and y as z changes. Then $\dfrac{y}{x} = \dfrac{f(z)}{\phi(z)}$. Since by hypothesis x and y become zero for some value of z, the curve MN passes through the origin. Let a be the value of z which renders x and y zero. Then as z approaches a, x and y approach zero, and P approaches O, so that the value of $\dfrac{y}{x}$ when $z = a$ is the limit of

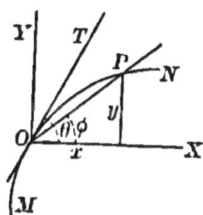

Fig. 11.

$\tan \phi$, ϕ being the angle which the secant makes with X. But the limit of $\tan \phi$ as P approaches O is $\tan \theta$, OT being the tangent at O; hence

$$\left.\frac{y}{x}\right]_{z=a} = \tan \theta = \left.\frac{dy}{dx}\right]_{z=a},$$

or

$$\left.\frac{f(z)}{\phi(z)}\right]_{z=a} = \left.\frac{f'(z)}{\phi'(z)}\right]_{z=a}.$$

Therefore, to find the value of $\dfrac{f(a)}{\phi(a)}$, we find that of $\dfrac{f'(a)}{\phi'(a)}$, since these are equal.

If $\dfrac{f'(a)}{\phi'(a)}$ is also $\dfrac{0}{0}$, then since $f'(z)$ and $\phi'(z)$ may be regarded as new functions of z whose ratio is $\dfrac{0}{0}$ when $z = a$,

$$\frac{f'(a)}{\phi'(a)} = \frac{f''(a)}{\phi''(a)},$$

and so on indefinitely. Hence

To evaluate a function which assumes the form $\frac{0}{0}$ for a particular value of the variable, form the successive derivatives of its numerator and denominator and substitute in them the particular value of the variable, continuing the process till a pair is found whose ratio does not become $\frac{0}{0}$.

EXAMPLES. Find the value of:

1. $\dfrac{\sin x}{x}$ when $x = 0$.

$$\frac{f'(x)}{\phi'(x)} = \frac{\cos x}{1}\bigg]_0 = 1.$$

2. $\dfrac{1 - \cos x}{x^2}$ when $x = 0$.

$$\frac{f'(x)}{\phi'(x)} = \frac{\sin x}{2x}\bigg]_0 = \frac{0}{0}; \quad \frac{f''(x)}{\phi''(x)} = \frac{\cos x}{2}\bigg]_0 = \frac{1}{2}.$$

3. $\dfrac{e^x - e^{-x}}{\log(1 + x)}$ when $x = 0$.   *Ans.* 2.

4. $\dfrac{a^x - b^x}{x}$ when $x = 0$.   *Ans.* $\log\dfrac{a}{b}$.

5. $\dfrac{x - 1}{x^n - 1}$ when $x = 1$.   *Ans.* $\dfrac{1}{n}$.

6. $\dfrac{x^3 - 5x^2 + 7x - 3}{x^3 - x^2 - 5x - 3}$ when $x = 3$.   *Ans.* $\frac{1}{4}$.

The successive differentiations will be facilitated by evaluating a factor in any result when possible. Thus:

7. $\dfrac{x - \sin^{-1} x}{\sin^3 x}$ when $x = 0$.

$$\frac{f'(x)}{\phi'(x)} = \frac{\sqrt{1 - x^2} - 1}{3 \sin^2 x \cos x \sqrt{1 - x^2}} = \frac{1}{\cos x \sqrt{1 - x^2}} \cdot \frac{\sqrt{1 - x^2} - 1}{3 \sin^2 x},$$

the first factor of which becomes 1 when $x = 0$. Proceeding with the second factor, $\dfrac{f''(x)}{\phi''(x)} = -\dfrac{1}{6 \cos x \sqrt{1 - x^2}} \dfrac{x}{\sin x}$, the first factor becoming $-\frac{1}{6}$ when $x = 0$. From Ex. 1 the value of the second factor when $x = 0$ is 1. Hence

$$\frac{x - \sin^{-1} x}{\sin^3 x} \bigg]_0 = -\frac{1}{6}.$$

8. $\dfrac{x^{\frac{3}{2}} - 1 + (x - 1)^{\frac{3}{2}}}{\sqrt{x^2 - 1}}$ when $x = 1$.   *Ans.* 0.

9. $\dfrac{x^3 - 3x + 2}{x^4 - 6x^2 + 8x - 3}$ when $x = 1$.   *Ans.* ∞.

10. $\dfrac{a^{\sin x} - a}{\log \sin x}$ when $x = \dfrac{\pi}{2}$.   *Ans.* $a \log a$.

11. $\dfrac{\sqrt{x} \tan x}{(e^x - 1)^{\frac{3}{2}}}$ when $x = 0$.   *Ans.* 1.

Write in the form $\sqrt{\dfrac{x}{e^x - 1} \dfrac{\tan x}{x} \dfrac{x}{e^x - 1}}$ and evaluate the factors separately.

12. $\dfrac{\tan x - \sin x}{x^3}$ when $x = 0$.   *Ans.* $\frac{1}{2}$.

13. $\dfrac{\tan x - x}{x - \sin x}$ when $x = 0$.   *Ans.* 2.

14. $\dfrac{e^x - e^{-x} - 2x}{(e^x - 1)^3}$ when $x = 0$.   *Ans.* $\frac{1}{3}$.

75. The form $\frac{\infty}{\infty}$. When $f(x)$ and $\phi(x)$ both increase indefinitely as x approaches a, then $\dfrac{f(x)}{\phi(x)}\bigg]_{x=a} = \dfrac{\infty}{\infty}$.

I. $\dfrac{f(x)}{\phi(x)} = \dfrac{\dfrac{1}{\phi(x)}}{\dfrac{1}{f(x)}}\Bigg]_{x=a} = \dfrac{0}{0}$. Hence the form $\frac{\infty}{\infty}$ can be reduced to

the form $\frac{0}{0}$ and treated as already explained. Thus

$$\frac{\sec x}{\sec 3x}\bigg]_{\frac{\pi}{2}} = \frac{\infty}{\infty}. \text{ But } \frac{\sec x}{\sec 3x} = \frac{\dfrac{1}{\cos x}}{\dfrac{1}{\cos 3x}} = \frac{\cos 3x}{\cos x}\bigg]_{\frac{\pi}{2}} = \frac{0}{0}.$$

Hence, by the process already established,

$$\frac{f'(x)}{\phi'(x)} = \frac{-3\sin 3x}{-\sin x}\bigg]_{\frac{\pi}{2}} = -3.$$

This transformation, however, will not always be successful unless the terms become infinite because of a denominator in each which becomes zero. Thus, in the above example, $\sec x$ becomes infinity, because it may be written $\dfrac{1}{\cos x}$ whose denominator becomes zero.

II. Since $\dfrac{f(x)}{\phi(x)} = \dfrac{\dfrac{1}{\phi(x)}}{\dfrac{1}{f(x)}}\Bigg]_{x=a} = \dfrac{0}{0}$, if we treat the latter by the

process of Art. 74, we have, when $x = a$,

$$\frac{f(x)}{\phi(x)} = \frac{-\dfrac{\phi'(x)}{[\phi(x)]^2}}{-\dfrac{f'(x)}{[f(x)]^2}} = \left[\frac{f(x)}{\phi(x)}\right]^2 \frac{\phi'(x)}{f'(x)}, \tag{1}$$

or $\quad 1 = \dfrac{f(x)}{\phi(x)} \dfrac{\phi'(x)}{f'(x)},$ \hfill (2)

whence
$$\frac{f(x)}{\phi(x)} = \frac{f'(x)}{\phi'(x)};$$

and the form $\frac{\infty}{\infty}$ can be treated *directly* in the same way as the form $\frac{0}{0}$.

Since all the derivatives of a function which becomes ∞ for a *finite* value of the variable also become infinite (Art. 56), this process would appear to lead to no result except when the given value of the variable is infinite, $\frac{f'(x)}{\phi'(x)}, \frac{f''(x)}{\phi''(x)}$, etc., becoming in turn $\frac{\infty}{\infty}$. This is true, but $\frac{f'(x)}{\phi'(x)}$ may, *by changing its form*, be more easily evaluated than $\frac{f(x)}{\phi(x)}$. Thus

$$\frac{\log x}{\frac{1}{x}}\Bigg]_0 = \frac{\infty}{\infty}.$$

$$\frac{f'(x)}{\phi'(x)} = \frac{\frac{1}{x}}{-\frac{1}{x^2}},$$ which also becomes $\frac{\infty}{\infty}$ for $x = 0$, but it may readily be put under the form $-\frac{x^2}{x} = -x]_0 = 0.$

In any case, therefore, when a function assumes the form $\frac{\infty}{\infty}$ for a finite value of the variable, it is necessary to transform either the function (I.), or some one of the derived functions (II.) so that it will not assume this form for the given value of the variable.

III. If the true value of $\frac{f(x)}{\phi(x)}$ when $x = a$ is zero or infinity, equation (1) is satisfied independently of equation (2), and it would therefore appear that in such cases the latter is not necessarily true. That equation (2) holds, however, when the true value of $\frac{f(x)}{\phi(x)}$ is zero or infinity may be shown as follows:

First. Let $\dfrac{f(x)}{\phi(x)} = 0$ when $x = a$. Then, if c be any finite quantity, $\dfrac{f(a)}{\phi(a)} + c$ is finite, and to this function the process of II. applies since it holds whenever the function does not become zero or infinity. Hence

$$\frac{f(a)}{\phi(a)} + c = \frac{f(a) + c\phi(a)}{\phi(a)} = \frac{f'(a) + c\phi'(a)}{\phi'(a)} = \frac{f'(a)}{\phi'(a)} + c,$$

or $\dfrac{f'(a)}{\phi'(a)} = 0$ when $\dfrac{f(a)}{\phi(a)} = 0$, and the process therefore gives the true value.

Second. Let $\dfrac{f(x)}{\phi(x)} = \infty$ when $x = a$. Then $\dfrac{\phi(a)}{f(a)} = 0$. Hence, by the preceding, $\dfrac{\phi(a)}{f(a)} = \dfrac{\phi'(a)}{f'(a)}$, or $\dfrac{f(a)}{\phi(a)} = \dfrac{f'(a)}{\phi'(a)}$, and the process holds in this case also.

EXAMPLES. Evaluate :

1. $\dfrac{\tan x}{\tan 5x}$ when $x = \dfrac{\pi}{2}$.

$$\frac{\tan x}{\tan 5x} = \frac{\dfrac{\sin x}{\cos x}}{\dfrac{\sin 5x}{\cos 5x}} = \frac{\sin x}{\sin 5x}\,\frac{\cos 5x}{\cos x}.$$

When $x = \dfrac{\pi}{2}$, the first factor is 1, and the second factor becomes $\dfrac{0}{0}$. Evaluating the latter by Art. 74, we find

$$\left.\frac{\tan x}{\tan 5x}\right]_{\frac{\pi}{2}} = 5.$$

2. $\dfrac{\sec \dfrac{\pi x}{2}}{\log (1 - x)}$ when $x = 1$.

$$\frac{f'(x)}{\phi'(x)} = \frac{\dfrac{\pi}{2}\sec\dfrac{\pi x}{2}\tan\dfrac{\pi x}{2}}{-\dfrac{1}{1-x}} = \frac{\tan\dfrac{\pi x}{2}}{-\dfrac{\cos\dfrac{\pi x}{2}}{1-x}}.$$

When $x = 1$, the denominator becomes $\frac{0}{0}$, and differentiating once we find its value to be -1. Hence

$$\frac{\sec \frac{\pi x}{2}}{\log (1-x)}\bigg]_1 = \infty.$$

3. $\dfrac{\log x}{x^n}$ when $x = \infty$.   *Ans.* 0, or ∞, as $n > 0$ or $n < 0$.

4. $\dfrac{\log \tan nx}{\log \tan x}$ when $x = 0$.   *Ans.* 1.

76. The form $0 \times \infty$. When, for $x = a$, $f(x) = 0$ and $\phi(x) = \infty$,

$$f(x)\,\phi(x) = f(x)\frac{1}{\dfrac{1}{\phi(x)}} = \frac{0}{0}, \text{ or } f(x)\,\phi(x) = \frac{1}{\dfrac{1}{f(x)}}\phi(x) = \frac{\infty}{\infty}.$$

Hence, by introducing the reciprocal of one of the factors, the function may be reduced to one of the two forms $\frac{0}{0}$, $\frac{\infty}{\infty}$, as is most convenient, and treated as before.

EXAMPLES. Evaluate:

1. $(1-x)\tan\dfrac{\pi x}{2}$ when $x = 1$.

$$(1-x)\tan\frac{\pi x}{2} = \frac{1-x}{\cot\dfrac{\pi x}{2}}\bigg]_1 = \frac{0}{0}.$$

Hence, by Art. 74,

$$\frac{1-x}{\cot\dfrac{\pi x}{2}}\bigg]_1 = \frac{-1}{-\dfrac{\pi}{2}\cosec^2\dfrac{\pi x}{2}}\bigg]_1 = \frac{2}{\pi}.$$

2. $e^{-\frac{1}{x}}(1 - \log x)$ when $x = 0$.

$$e^{-\frac{1}{x}}(1 - \log x) = \frac{1 - \log x}{e^{\frac{1}{x}}}\bigg]_0 = \frac{\infty}{\infty}.$$

Hence, by Art. 75,

$$\frac{1 - \log x}{e^{\frac{1}{x}}}\Big]_0 = 0.$$

3. $e^x \sin \dfrac{a}{e^x}$ when $x = \infty$.

$$e^x \sin \frac{a}{e^x} = \frac{\sin \dfrac{a}{e^x}}{e^{-x}}\Big]_{\infty} = \frac{0}{0}.$$
Ans. a.

4. $(a^{\frac{1}{x}} - 1)x$ when $x = \infty$.   Ans. $\log a$.

77. The form $\infty - \infty$. When, for $x = a$,

$$f(x) = \infty \text{ and } \phi(x) = \infty,$$

$$f(x) - \phi(x) = \frac{1}{\dfrac{1}{f(x)}} - \frac{1}{\dfrac{1}{\phi(x)}} = \frac{\dfrac{1}{\phi(x)} - \dfrac{1}{f(x)}}{\dfrac{1}{f(x)\phi(x)}} = \frac{0}{0},$$

and may be treated when thus transformed by Art. 74.

EXAMPLES. Evaluate:

1. $\dfrac{1}{\log x} - \dfrac{x}{\log x}$ when $x = 1$.

Here $f(x) = \dfrac{1}{\log x}$, $\phi(x) = \dfrac{x}{\log x}$.

Hence $\dfrac{\dfrac{1}{\phi(x)} - \dfrac{1}{f(x)}}{\dfrac{1}{f(x)\phi(x)}} = \dfrac{\dfrac{\log x}{x} - \log x}{\dfrac{1}{\dfrac{x}{\log^2 x}}} = \dfrac{1 - x}{\log x}\Big]_1 = \dfrac{0}{0},$

and, by Art. 74,

$$\frac{1 - x}{\log x}\Big]_1 = \frac{-1}{\dfrac{1}{x}}\Big]_1 = -1.$$

2. $\dfrac{x}{x-1} - \dfrac{1}{\log x}$ when $x = 1$.

Transforming, as above, we obtain

$$\left.\frac{x \log x - x + 1}{x \log x - \log x}\right]_1 = \frac{0}{0}.$$

Hence $\left.\dfrac{x \log x - x + 1}{x \log x - \log x}\right]_1 = \left.\dfrac{\log x}{\log x + 1 - \dfrac{1}{x}}\right]_1 = \left.\dfrac{x}{x+1}\right]_1 = \dfrac{1}{2}.$

3. $\sec x - \tan x$ when $x = \dfrac{\pi}{2}$.

This may be transformed as above; or, more directly,

$$\sec x - \tan x = \frac{1}{\cos x} - \frac{\sin x}{\cos x} = \left.\frac{1 - \sin x}{\cos x}\right]_{\frac{\pi}{2}} = \frac{0}{0}.$$

$Ans.\ 0.$

4. $\dfrac{2}{x^2 - 1} - \dfrac{1}{x-1}$ when $x = 1$.

This may be transformed as above; or, reducing to a common denominator,

$$\frac{2}{x^2 - 1} - \frac{1}{x - 1} = \left.\frac{2x - x^2 - 1}{x^3 - x^2 - x + 1}\right]_1 = \left.\frac{2 - 2x}{3x^2 - 2x - 1}\right]_1$$

$$= \left.\frac{-2}{6x - 2}\right]_1 = -\frac{1}{2}.$$

5. $x \tan x - \dfrac{\pi}{2} \sec x$ when $x = \dfrac{\pi}{2}$. $\qquad\qquad Ans.\ -1.$

78. The forms ∞^0, 1^∞, 0^0. The logarithm of a power may assume an illusory form under the following circumstances. Let z^y be the function. Passing to logarithms,

$$\log z^y = y \log z,$$

which becomes $0 \times \infty$ when $\begin{cases} y = 0 \text{ and } z = 0, & \therefore z^y = 0^0 \\ y = 0 \text{ and } z = \infty, & \therefore z^y = \infty^0, \end{cases}$

and which becomes $\infty \times 0$ when $y = \infty$ and $z = 1, \ \therefore z^y = 1^\infty.$

The logarithms of such functions may therefore be evaluated

as in Art. 76, and thus the values of the functions themselves are readily obtained.

The functions 0^∞ and ∞^∞ do not give rise to illusory forms, as may be seen by passing to logarithms, the logarithms in both cases being infinity.

EXAMPLES. Evaluate:

1. $\left(\dfrac{a}{x}+1\right)^x$ when $x=\infty$.

Putting $v=\left(\dfrac{a}{x}+1\right)^x$, $\log v = x\log\left(\dfrac{a}{x}+1\right)=\left.\dfrac{\log\left(\dfrac{a}{x}+1\right)}{\dfrac{1}{x}}\right]_\infty$

$$=\left.\dfrac{-\dfrac{a}{x^2}\dfrac{x}{a+x}}{-\dfrac{1}{x^2}}=\dfrac{ax}{a+x}\right]_\infty=\dfrac{a}{1}, \quad \therefore v=e^a.$$

2. $\left(1+\dfrac{1}{x^2}\right)^x = v$, when $x=\infty$. *Ans.* $\log v = 0,\ \therefore v = e^0 = 1$.

3. $(\sin x)^{\tan x}$ when $x=\dfrac{\pi}{2}$. *Ans.* 1.

4. x^x when $x=0$.

$$v=x^x\cdot \log v = x\log x = \left.\dfrac{\log x}{\dfrac{1}{x}}\right]_0 = \left.\dfrac{\dfrac{1}{x}}{-\dfrac{1}{x^2}}=-x\right]_0=0.$$

Hence $v=1$.

5. $x^{\frac{1}{1-x}}$ when $x=1$.

$$\log v = \left.\dfrac{\log x}{1-x}\right]_1 = \dfrac{\dfrac{1}{x}}{-1}=\left.-\dfrac{1}{x}\right]_1=-1,\ \therefore v=\dfrac{1}{e}.$$

6. $(\sin x)^{\sin x}$ when $x=0$. *Ans.* 1.

7. $(\cot x)^{\sin x}$ when $x=0$.

$$(\cot x)^{\sin x}=\left.\dfrac{(\cos x)^{\sin x}}{(\sin x)^{\sin x}}\right]_0 = \dfrac{1}{0^0},\ \text{or (Ex. 6), 1.}$$

8. $(\sin x)^{\tan x}$ when $x = 0$.   *Ans.* 1.

9. $(1 + ax)^{\frac{1}{x}}$ when $x = 0$.   *Ans.* e^a.

10. $x^{e^x - 1}$ when $x = 0$.   *Ans.* 1.

Maxima and Minima Values of a Function of a Single Variable.

79. *The value of a function is said to be a* **maximum** *when it is greater than its immediately preceding and succeeding values, and a* **minimum** *when it is less than its immediately preceding and succeeding values.*

By greater and less values are meant algebraic values. Thus, if MN be the locus of $y = f(x)$, and mn is greater than the immediately preceding and succeeding ordinates, mn is a maximum value of y. Similarly pq is a minimum value of y. It is evident that for increasing values of x, y diminishes after passing through a maximum value, and cannot therefore have a second maximum value without first passing through a minimum; or maxima and minima values occur alternately. From the definition it is also evident that a maximum value is not the greatest possible value, nor a minimum the least possible value, of a function.

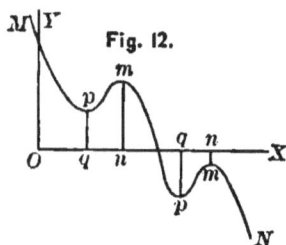

Fig. 12.

80. Condition of a maximum or minimum value.

For increasing values of x, $f(x)$ is increasing before, and decreasing after, a maximum value. Hence (Art. 22), $f'(x)$ is positive before, and negative after, a maximum value of $f(x)$; or the first derivative of the function changes sign from plus to minus as the function passes through a maximum value.

Similarly, a function decreases as it approaches a minimum value and increases after such value; or the first derivative changes sign from minus to plus as the function passes through a minimum value.

Since in either case the first derivative changes sign, it must pass through zero or infinity. Hence, *every value of x which renders f(x) a maximum or a minimum is a root of $f'(x) = 0$, or of $f'(x) = \infty$.*

It is to be observed that the essential characteristic of a maximum or minimum value of the function is *a change of sign of its first derivative.* Now a quantity may become zero or infinity without changing sign; hence the roots of $f'(x) = 0$ and $f'(x) = \infty$ are called **critical** values, and must be separately examined; only those for which $f'(x)$ changes sign can correspond to maxima or minima values of the function.

81. Geometric illustrations. Since $y = f(x)$ is the equation of some locus, and $f'(x)$ is the slope of the locus at any point, the foregoing remarks admit of the following illustration:

In Fig. 13, Pm being a maximum value of y, for increasing values of x the angle made by the tangent with X is acute before, and becomes obtuse after, the maximum value; hence the tangent of this angle, which is $f'(x)$, is positive before and negative after this value. At P the tangent is parallel to X, and its slope is therefore zero.

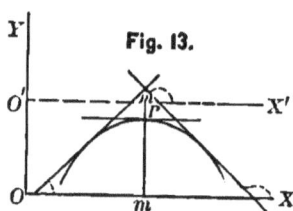

Fig. 13.

In Fig. 14, Pm being a minimum value of y, the angle made by the tangent with X is obtuse before and acute after the minimum value of y, the slope changing from minus to plus, and passing through zero as before.

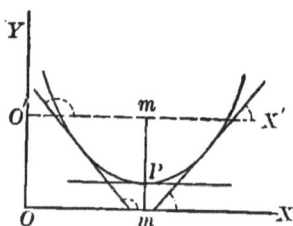

Fig. 14.

In Fig. 15, although the tangent at P is parallel to X and therefore $f'(x)$ is then zero, the angle is obtuse both before and after the value $x = Om$ and does not change sign; hence Pm is

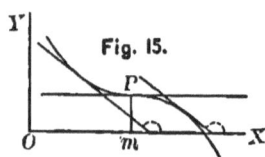

Fig. 15.

neither a maximum nor a minimum value of y.

The change of sign of $f'(x)$ from $+$ to $-$, and from $-$ to $+$, in passing through infinity is shown in Fig. 16, the tangent at P being perpendicular to X and its slope infinity.

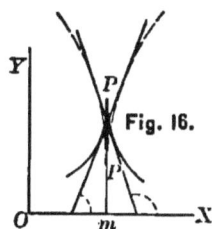

Fig. 16.

82. Examination of the critical values when $f'(x) = 0$.

Since $f'(x)$ changes sign from $+$ to $-$ as $f(x)$ passes through a maximum value, it is a decreasing function, and its first derivative $f''(x)$ must be negative (Art. 22).

Also, since $f'(x)$ changes sign from $-$ to $+$ as $f(x)$ passes through a minimum value, it is an increasing function, and its first derivative $f''(x)$ must be positive.

Hence, *to examine $f(x)$ for maxima or minima values, observe whether $f''(x)$ is negative or positive for critical values of x,* that is, for values derived from the equation $f'(x) = 0$.

As the second derivative may become zero for a critical value of x, the above test may fail. To provide for such case we have the following more general rule.

83. Let $y = f(x)$, and $y_1 = f(x_1)$ in which x_1 is the value of x which renders $y = y_1 = $ a maximum or a minimum.

Let $y' = f(x_1 - h)$ and $y'' = f(x_1 + h)$ be the values immediately preceding and succeeding the maximum or minimum value y_1, $x_1 - h$ and $x_1 + h$ being the corresponding values of x. Developing y' and y'' by Taylor's formula, we have

$$y' = f(x_1 - h) = f(x_1) - f'(x_1)h + f''(x_1)\frac{h^2}{\lfloor 2}$$

$$-f'''(x_1)\frac{h^3}{\lfloor 3} + f^{iv}(x_1)\frac{h^4}{4} \cdots$$

$$y'' = f(x_1 + h) = f(x_1) + f'(x_1)h + f''(x_1)\frac{h^2}{\lfloor 2}$$

$$+f'''(x_1)\frac{h^3}{\lfloor 3} + f^{iv}(x_1)\frac{h^4}{4} \cdots$$

But $f(x)_1 = y_1$, and since x_1 corresponds to a maximum or a minimum, $f'(x_1) = 0$. Hence, transposing,

$$y' - y_1 = f''(x_1)\frac{h^2}{\lfloor 2} - f'''(x_1)\frac{h^3}{\lfloor 3} + f^{\text{iv}}(x_1)\frac{h^4}{\lfloor 4} \cdots, \qquad (1)$$

$$y'' - y_1 = f''(x_1)\frac{h^2}{\lfloor 2} + f'''(x_1)\frac{h^3}{\lfloor 3} + f^{\text{iv}}(x_1)\frac{h^4}{\lfloor 4} \cdots. \qquad (2)$$

Now the signs of the second members of (1) and (2) will be those of their first terms, that is of $f''(x_1)$, if h be taken sufficiently small; and since h approaches zero as the function approaches its maximum or minimum, we are at liberty to make h as small as we please. Hence if $f''(x_1)$ is positive, the first members are positive, and both y' and y'' greater than y_1, which is therefore a minimum; while if $f''(x_1)$ is negative, the first members are negative, both y' and y'' are less than y_1, and y_1 is a maximum. This accords with what has already been said.

If $f''(x_1)$ is zero, then

$$y' - y_1 = -f'''(x_1)\frac{h^3}{\lfloor 3} + f^{\text{iv}}(x_1)\frac{h^4}{\lfloor 4} \cdots,$$

$$y'' - y_1 = f'''(x_1)\frac{h^3}{\lfloor 3} + f^{\text{iv}}(x_1)\frac{h^4}{\lfloor 4} \cdots,$$

in which, whatever the sign of $f'''(x_1)$, the first members will have opposite signs, and y' and y'' cannot both be greater than y_1, nor both less. Hence neither a maximum nor a minimum can exist unless $f'''(x_1) = 0$. If this condition be fulfilled, there will be a maximum or a minimum according as $f^{\text{iv}}(x_1)$ is negative or positive. We have therefore the following rule:

To determine whether a function has maxima or minima values, form its first derivative and place it equal to zero. The roots of this equation contain the values of the variable which correspond to either maxima or minima values of the function. Find the first derivative which does not become zero for one of these critical values of the variable. If this derivative is of an odd order, there is neither a maximum nor a minimum; if of an even order, the function is a maximum or a minimum according as the derivative is negative or positive.

Each critical value must of course be examined in turn.

ILLUSTRATION. Examine $x^5 - 5x^4 + 5x^3 + 1$ for maxima and minima values.

$$f'(x) = 5x^4 - 20x^3 + 15x^2 = 5x^2(x^2 - 4x + 3) = 0.$$

The roots of this equation are $x = 0$, $x = 1$, $x = 3$.

$$f''(x) = 20x^3 - 60x^2 + 30x = 10x(2x^2 - 6x + 3).$$

Substituting $x = 3$, $f''(x) = +90$; hence $x = 3$ renders the function a minimum, and substituting this value of x in the function we find $f(x) = -26$, the minimum.

Substituting $x = 1$, $f''(x) = -10$; hence $x = 1$ renders the function a maximum, which we find to be 2.

As $f''(x) = 0$ for $x = 0$, we form $f'''(x) = 60x^2 - 120x + 30$, which does not vanish for $x = 0$ and is of an odd order. Hence $x = 0$ corresponds to neither a maximum nor a minimum.

84. Abbreviated processes.

I. Since the essential characteristic of a maximum or minimum value of a function is a change in the sign of its first derivative, it will be sufficient, when possible, to observe whether for a critical value of the variable such change actually takes place. Thus, from $(x - a)^4 + b$, $f'(x) = 4(x - a)^3 = 0$, the critical value being $x = a$. Now in passing through $x = a$, $f'(x)$ changes sign from $-$ to $+$; hence $x = a$ renders the function a minimum, namely b. Again, from $(x - a)^3 + b$, $f'(x) = 3(x - a)^2 = 0$, which cannot change sign for any value of x; hence the function has no maxima nor minima values.

II. Since if A is a constant factor, $Af(x)$ increases and decreases with $f(x)$, a constant factor may be omitted in the search for maxima or minima values.

III. Since $\pm A + f(x)$ increases and decreases with $f(x)$, we may substitute $f(x)$ for $\pm A + f(x)$ in searching for maxima or minima values. If $A - f(x)$ is the given function, we may substitute $f(x)$, provided we reverse the conclusions, as $A - f(x)$ increases when $f(x)$ decreases, and decreases when $f(x)$ increases.

IV. Since $\dfrac{1}{f(x)}$ decreases as $f(x)$ increases, and conversely, the reciprocal of the function may be substituted for the function, provided the conclusions are reversed.

V. Since $\log [f(x)]$ increases and decreases with $f(x)$, the number may be substituted for the logarithm of the number in the search for maxima and minima values.

VI. If $f(x)$ is positive, $[f(x)]^n$ is also positive, and therefore increases and decreases with $f(x)$; or any power of a positive function may be substituted for the function. If $f(x)$ is negative, $[f(x)]^n$ will have the same sign as $f(x)$ if n is odd, but the opposite sign if n is even; or any power of a negative function may be substituted for the function, provided the conclusions are reversed if n is even.

We are thus enabled to omit the radical sign in the search for maxima and minima values of any positive radical; also when the radical is negative, if we reverse the conclusions.

EXAMPLES. Examine the following functions for maxima and minima values.

1. $x^3 - 9x^2 + 15x - 3$.

Omitting the constant term (III., Art. 84),

$$f(x) = x^3 - 9x^2 + 15x.$$

$f'(x) = 3x^2 - 18x + 15 = 0$, whence the critical values

$$x = 5,\ x = 1.$$

$f''(x) = 6x - 18$, which is 12 for $x = 5$ and -12 for $x = 1$. Hence $x = 5$ renders the function a minimum, and $x = 1$ renders it a maximum. Substituting $x = 5$ and $x = 1$ in the function, its minimum and maximum are found to be -28 and 4, respectively.

2. $b + c(x - a)^{\frac{4}{3}}$.

$f(x) = (x - a)^{\frac{4}{3}}$ (Art. 84, II. and III.).

$f'(x) = \frac{4}{3}(x - a)^{\frac{1}{3}} = 0$, whence $x = a$; and as $f'(x)$ changes sign from $-$ to $+$ for increasing values of x as x passes through a, b is a minimum value of the function.

3. $x^5 - 5x^4 + 5x^3 - 6$.

4. Examine the circle $y^2 + x^2 = R^2$ for maxima and minima ordinates.

The function to be examined is $y = \pm \sqrt{R^2 - x^2}$. Omitting the radical (Art. 84, VI.), $f(x) = R^2 - x^2$, whence $f'(x) = -2x = 0$, or $x = 0$; and as $f'(x)$ changes sign from $+$ to $-$ as x passes through 0, $x = 0$ corresponds to a maximum. If we take the negative value of the function, then, in omitting the radical, we raise the function to an even power and must reverse the conclusion; hence when y is negative, $x = 0$ corresponds to a minimum.

5. $(x - 1)^4 (x + 2)^3$.

$f'(x) = 4(x - 1)^3 (x + 2)^3 + 3(x - 1)^4 (x + 2)^2$

$\qquad = (x - 1)^3 (x + 2)^2 (7x + 5)$; whence the critical values $x = 1$, $x = -2$, $x = -\frac{5}{7}$.

Since $f'(x)$ is $-$ if x is a little less than 1, and $+$ if x is a little greater than 1, it changes sign from $-$ to $+$ as x passes through 1; hence $x = 1$ corresponds to a minimum.

$f''(x) = 3(x-1)^2 (x+2)^2 (7x+5) + 2(x-1)^3 (x+2)(7x+5)$
$$+ 7(x-1)^3 (x+2)^2$$

$\qquad = (x-1)^2 (x+2)\{3(x+2)(7x+5) + 2(x-1)(7x+5)$
$$+ 7(x-1)(x+2)\}$$

$\qquad = 6(x-1)^2 (x+2)(7x^2 + 10x + 1)$.

When $x = -\frac{5}{7}$, the first two factors are positive, and the sign will depend upon that of the third factor, which is $-$ for $x = -\frac{5}{7}$; hence $x = -\frac{5}{7}$ corresponds to a maximum.

Since $f''(x) = 0$ for $x = -2$, $f'''(x) = 6(x-1)^2(7x^2 + 10x + 1) +$ other terms which contain $(x+2)$, and which therefore vanish when $x = -2$, while the term $6(x-1)^2(7x^2 + 10x + 1)$ does not. Hence $f'''(x)$ does not become zero for $x = -2$, and this value of x corresponds to neither a minimum nor a maximum.

6. $x^3 - 3x^2 + 6x + 7$. The critical values are imaginary.

7. $\text{Sin}^3 x \cos x$.

$f'(x) = 3 \sin^2 x \cos^2 x - \sin^4 x = 3 \sin^2 x(1 - \sin^2 x) - \sin^4 x$
$$= 3 \sin^2 x - 4 \sin^4 x = 0 \; ; \text{ whence}$$

$\sin^2 x(3 - 4 \sin^2 x) = 0$, and the critical values are $\sin x = 0$, $\sin x = \frac{\sqrt{3}}{2}$, or $x = 0°$, $x = 60°$. Since $f'(x)$ evidently changes sign from $+$ to $-$ as $\sin x$ passes through the value $\frac{\sqrt{3}}{2}$, $x = 60°$ corresponds to a maximum. If x is a little greater or less than $0°$, $4 \sin^2 x < 3$ and $f'(x)$ is $+$; hence $x = 0°$ corresponds to neither a maximum nor a minimum.

8. $a + \sqrt{4x^2 - 2x^3}$.

Omitting the constant term, radical sign, and factor 2 (Art. 84, III., VI., II.), we have $2x^2 - x^3$; whence $f'(x) = 4x - 3x^2 = 0$, or $x = 0$, $x = \frac{4}{3}$.

$f''(x) = 4 - 6x$, which is $+$ for $x = 0$ and $-$ for $x = \frac{4}{3}$. Hence the function is a minimum when $x = 0$ and a maximum when $x = \frac{4}{3}$.

9. Divide a into two factors, the sum of which shall be a minimum.

Let $x =$ one factor ; then $\frac{a}{x} =$ the other, and the function is $x + \frac{a}{x}$; or $f'(x) = 1 - \frac{a}{x^2} = 0$, whence $x = \sqrt{a}$, and the factors are equal.

10. The difference between two members is a. Prove that the greater $=$ twice the less when the square of the greater divided by the less is a minimum.

11. Find a number x such that its xth root shall be a maximum.   *Ans.* $x = e$.

12. To determine the number of equal parts into which a must be divided in order that their continued product may be a maximum.

Let $x =$ number of parts; then $\dfrac{a}{x} =$ one part, and $\dfrac{a}{x} \cdot \dfrac{a}{x} \cdot \dfrac{a}{x} \cdots$ to x factors $= \left(\dfrac{a}{x}\right)^x$ is to be a maximum.

$$\log\left(\dfrac{a}{x}\right)^x = x(\log a - \log x) = f(x).$$

$f'(x) = \log a - \log x - 1 = 0$, or $\log \dfrac{a}{x} = 1$, whence $\dfrac{a}{x} = e$, or $x = \dfrac{a}{e}.$

Arithmetically the problem would not be possible unless a was a multiple of e, otherwise x would not be an integer. The general solution belongs to the statement: to find a number x such that the xth power of $\dfrac{a}{x}$ shall be a maximum.

13. $\dfrac{x}{1 + x \tan x} \cdot f'(x) = \dfrac{1 - \dfrac{x^2}{\cos^2 x}}{(1 + x \tan x)^2}.$

A maximum when $x = \cos x.$

14. $\dfrac{\sin x}{1 + \tan x} \cdot f'(x) = \dfrac{1 - \tan^3 x}{\dfrac{(1 + \tan x)^2}{\cos x}}.$

A maximum when $x = 45°.$

85. Examination of the critical values when $f'(x) = \infty$.

Since, when $f'(x) = \infty$ for a particular value of x, $f''(x)$, $f'''(x)$, etc., also become infinity (Art. 56), the function cannot be developed by Taylor's formula, and the results of Art. 83 are inapplicable. In such cases we may examine the first derivative directly to see if it changes sign as the variable passes through its critical value.

EXAMPLES.   1. $b + (x - a)^{\frac{2}{3}}$.

$$f'(x) = \tfrac{2}{3}(x - a)^{-\frac{1}{3}} = \frac{2}{3(x - a)^{\frac{1}{3}}} = \infty, \text{ whence } x - a = 0, \text{ or}$$

$x = a$. It is readily seen that $f'(x)$ changes sign from $-$ to $+$, and that $x = a$ therefore corresponds to a minimum.

2. $\dfrac{(x + 2)^3}{(x - 3)^2}$.

$$f'(x) = \frac{(x+2)^2(x - 13)}{(x - 3)^3}.$$

$f'(x) = 0$ gives $x = -2$ and $x = 13$. $f'(x) = \infty$ gives $x = 3$. $x - 13$ is negative if x is a little less or greater than 3, while $(x - 3)^3$ is negative if $x < 3$ and positive if $x > 3$. Hence $f'(x)$ changes sign from $+$ to $-$ at $x = 3$, which gives a maximum.

$x = -2$ and $x = 13$ may be examined in like manner; the latter gives a minimum, and the former neither a maximum nor a minimum.

3. $\dfrac{(x - 1)^2}{(x + 1)^3}$.

$f'(x) = 0$ gives $x = 1$ and $x = 5$, the former corresponding to a minimum, and the latter to a maximum. $f'(x) = \infty$ gives $x = -1$, which corresponds to neither.

86. Geometrical Problems.

In the following problems $V =$ volume, $A =$ area, $S =$ surface, and the substituted function obtained after omitting constant factors, radical sign, etc. (Art. 84), is designated by an accent.

1. Determine the rectangle of greatest area which can be inscribed in a given circle.

Let $R =$ radius. If we take x, y, to represent the half-sides of the rectangle, then the equation of the circle gives the relation $x^2 + y^2 = R^2$, by means of which we can eliminate y from the expression for the area $A = 4xy$, obtaining

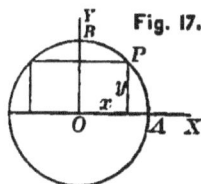

Fig. 17.

$$4x\sqrt{R^2 - x^2}, \text{ or } 4\sqrt{R^2 x^2 - x^4},$$

thus reducing the function to be examined to one of a single variable. Omitting the factor 4 and the radical, we have the substituted function $A' = R^2 x^2 - x^4$, whence

$$f'(x) = 2R^2 x - 4x^3 = 0, \text{ or } x = 0 \text{ and } x = \frac{R}{\sqrt{2}}.$$

$$f''(x) = 2R^2 - 12x^2, \text{ which becomes } -4R^2 \text{ for } x = \frac{R}{\sqrt{2}};$$

hence $\dfrac{R}{\sqrt{2}}$ corresponds to a maximum. Substituting $x = \dfrac{R}{\sqrt{2}}$

in $y^2 + x^2 = R^2$, we find $y = \dfrac{R}{\sqrt{2}}$; hence $x = y$, the rectangle is a square, and its area $A = 4xy = 2R^2$.

Before proceeding to the remaining examples the student will observe: 1°. As the point P moves from A to B, the area of the rectangle increases from 0, passes through its maximum, and decreases again to 0. Whenever, then, the conditions of the problem are such that the existence of a maximum value is clearly seen, it will be unnecessary to test the critical value. 2°. The solution consists in first finding an expression for the quantity to be examined, as $4xy$ in the above case. If this is a function of two variables, the next step is to eliminate one by means of some relation between them furnished by the conditions of the problem, as in the above case $y^2 + x^2 = R^2$. 3°. This elimination may be effected before or after differentiation. In the above case y was eliminated before finding the derivative; but we might have proceeded as follows:

$$A = 4xy; \quad A' = xy; \quad f'(x) = x\frac{dy}{dx} + y = 0.$$

From $x^2 + y^2 = R^2$, $\dfrac{dy}{dx} = -\dfrac{x}{y}$, hence $f'(x) = -x\dfrac{x}{y} + y = 0$, or $x^2 = y^2$, as before. Eliminating now y by substituting $y^2 = x^2$ in $x^2 + y^2 = R^2$, we have $x = \dfrac{R}{\sqrt{2}}$. It is frequently preferable thus to eliminate after differentiating.

2. Determine the rectangle of greatest area which can be inscribed in a given ellipse.

With the notation of Ex. 1, $A = 4xy$, the auxiliary relation being $a^2y^2 + b^2x^2 = a^2b^2$, the equation of the ellipse. Hence $A = 4\dfrac{b}{a}x\sqrt{a^2 - x^2}$, $A' = a^2x^2 - x^4$, $f'(x) = 2a^2x - 4x^3 = 0$, and $x = \dfrac{a}{\sqrt{2}}$, or $2x = a\sqrt{2}$, which, substituted in the equation of the ellipse, gives $2y = b\sqrt{2}$, the sides of the rectangle.

3. Determine the rectangle of greatest area which can be inscribed in a given segment of a parabola.

Let $OA = a$, and $y =$ the half-side AB. Then $A = 2y(a - x)$, or, since $y^2 = 2px$,
$$A = 2\sqrt{2px}(a - x) = 2\sqrt{2p}\sqrt{x(a - x)^2},$$
whence

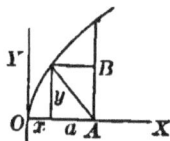
Fig. 18.

$$A' = a^2x - 2ax^2 + x^3, \quad f'(x) = a^2 - 4ax + 3x^2 = 0,$$

from which we find $x = \dfrac{a}{3}$. Therefore $a - x = \tfrac{2}{3}a =$ one side, and $2y = 2\sqrt{2px} = 2\sqrt{\dfrac{2pa}{3}} =$ the other side, and
$$A = 2y(a - x) = \dfrac{4}{3}\sqrt{\dfrac{2a^3p}{3}}.$$

4. Find the cylinder of greatest volume which can be inscribed in a given sphere.

With the notation of Ex. 1,
$$V = 2\pi y^2 x = 2\pi x(R^2 - x^2),$$

$$V' = R^2 x - x^3, \quad f'(x) = R^2 - 3 x^2 = 0,$$

whence $\quad x = \dfrac{R}{\sqrt{3}}, \text{ or } 2x = \dfrac{2R}{\sqrt{3}} = \text{altitude.}$

5. Find the cylinder of greatest convex surface which can be inscribed in a given sphere.

With the notation of Ex. 1,

$$S = 4\pi yx = 4\pi x \sqrt{R^2 - x^2},$$

$$S' = R^2 x^2 - x^4, \quad f'(x) = 2 R^2 x - 4 x^3 = 0,$$

whence $\quad x = \dfrac{R}{\sqrt{2}}, \text{ and } 2x = R\sqrt{2} = \text{altitude.}$

6. Find the cylinder of greatest volume which can be inscribed in a given ellipsoid.

With the notation of Ex. 1, using the equation of the ellipse,

$$V = 2\pi y^2 x = 2\pi \frac{b^2}{a^2} x (a^2 - x^2),$$

$$V' = a^2 x - x^3, \quad f'(x) = a^2 - 3 x^2 = 0,$$

whence $\quad x = \dfrac{a}{\sqrt{3}}, \text{ and } 2x = \dfrac{2a}{\sqrt{3}} = \text{altitude.}$

7. Find the cone of greatest volume which can be inscribed in a given sphere.

With the notation of the figure, $V = \frac{\pi}{3} y^2 x$;

but $y^2 = 2 Rx - x^2$, hence $V = \frac{\pi}{3}(2 Rx^2 - x^3)$,

$V' = 2 Rx^2 - x^3, \quad f'(x) = 4 Rx - 3 x^2 = 0$, or

$x = \frac{4}{3} R = \text{altitude.}$

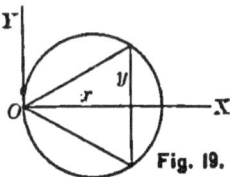

Fig. 19.

8. Find the cone of maximum convex surface which can be inscribed in a given sphere.

$$S = \pi y \sqrt{x^2 + y^2} = \pi \sqrt{2 Rx - x^2} \sqrt{2 Rx} = \pi \sqrt{4 R^2 x^2 - 2 Rx^3},$$

$$S' = 2 Rx^2 - x^3, \quad f'(x) = 4 Rx - 3 x^2 = 0,$$

and $x = \frac{4}{3} R = \text{altitude.}$

9. Find the cone of greatest volume which can be inscribed in a given paraboloid, the vertex of the cone being on the axis in the base of the paraboloid.

$$V = \frac{\pi}{3} y^2 (a - x). \quad \text{Altitude} = \frac{a}{2}. \quad \text{(Fig. 18.)}$$

10. Find the cylinder of greatest volume which can be inscribed in a given right cone.

Let b = radius of base, and a = altitude of the cone, and x, y, those of the cylinder. Then $V = \pi x^2 y$. From similar triangles, $b : a :: x : (a - y)$,

or $y = \frac{a}{b}(b - x)$.

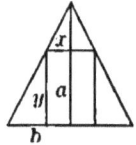
Fig. 20.

Hence　　$V = \frac{\pi a}{b} x^2 (b - x), \quad V' = bx^2 - x^3,$

$$f'(x) = 2bx - 3x^2 = 0, \quad \text{whence } x = \tfrac{2}{3}b,$$

and　　$y = \frac{a}{b}(b - x) = \frac{a}{3} = \text{altitude}.$

11. Find the cylinder of greatest convex surface which can be inscribed in a given right cone.

Ans. Altitude $= \frac{1}{2}$ altitude of cone.

12. Of all right cones of given convex surface determine that one whose volume is greatest.

If x = altitude, y = radius of base, $V = \frac{\pi}{3} y^2 x$. By condition, $\pi y \sqrt{x^2 + y^2} = m$, a constant. Differentiating first, from $V' = y^2 x$ we have $f'(x) = 2yx \frac{dy}{dx} + y^2 = 0$. From $\pi y \sqrt{x^2 y^2} = m$,

$\frac{dy}{dx} = -\frac{xy}{x^2 + 2y^2}$. Hence $f'(x) = -\frac{2 y^2 x^2}{x^2 + 2y^2} + y^2 = 0$, whence $x = y\sqrt{2}$, or the altitude $= \sqrt{2} \times$ radius of base.

13. Of all cones whose slant heights are equal find that which has the greatest volume.

Ans. The tangent of the semi-vertical angle $= \sqrt{2}$.

14. From a given quantity of material a cylindrical vessel with circular base and open top is to be made so as to have a maximum content. Find the relation between the radius and altitude.

Let $x =$ altitude, $y =$ radius. Then $V = \pi y^2 x$, $V' = y^2 x$,

$f'(x) = 2yx\dfrac{dy}{dx} + y^2 = 0$. By condition $2\pi yx + \pi y^2 = m$, a con-

stant; whence $\dfrac{dy}{dx} = -\dfrac{y}{x+y}$, and therefore

$$f'(x) = -\frac{2xy^2}{x+y} + y^2 = 0, \text{ or } x = y.$$

15. A square is cut from each corner of a rectangular piece of pasteboard whose sides are a and b. Find the side of the square that the remainder may form a box of maximum content.

$$\textit{Ans. The side} = \frac{a+b-\sqrt{a^2-ab+b^2}}{6}.$$

16. Prove that in an ellipse referred to its centre and axes, the product of the co-ordinates of a point on the curve is a maximum when the co-ordinates are in the ratio of the axes.

17. A vertical flagstaff consists of two pieces, the upper being a and the lower b feet long. Find the distance from the foot of the staff at which the visual angle subtended by the upper segment is a maximum.

With the notation of the figure,

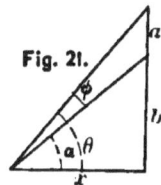

Fig. 21.

$$\tan \theta = \frac{a+b}{x}, \ \tan a = \frac{b}{x},$$

$$\tan \phi = \tan(\theta - a) = \frac{\tan \theta - \tan a}{1 + \tan \theta \tan a}$$

$$= \frac{\dfrac{a+b}{x} - \dfrac{b}{x}}{1 + \dfrac{ab+b^2}{x^2}}; \text{ whence } x = \sqrt{b(a+b)}.$$

18. Find the least triangle which can be circumscribed about a given ellipse having one side parallel to the transverse axis. *Ans.* Altitude $= 3b$, base $= 2a\sqrt{3}$.

19. Find the parabola of maximum area which can be cut from a given right cone, knowing the area of a parabola to be $\frac{2}{3} MQ \cdot QP$ (Fig. 22).

Let $AB = 2b$, $AC = a$, $QB = x$.

Then $MQ = \sqrt{AQ \cdot QB} = \sqrt{(2b - x)x}$,

and $AB:AC::QB:QP, QP = \dfrac{AC \cdot QB}{AB} = \dfrac{ax}{2b}$.

Hence $A = \frac{2}{3} MQ \cdot QP = \dfrac{2}{3}\dfrac{ax}{2b}\sqrt{(2b - x)x}$,

$A' = 2bx^3 - x^4$, $f'(x) = 6bx^2 - 4x^3 = 0$;

whence $x = QB = \frac{3}{2}b$. Hence $QP = \dfrac{ax}{2b}$

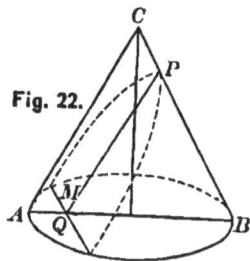

Fig. 22.

$= \frac{3}{4}a$, or the area of the parabola whose axis is $\frac{3}{4}$ the slant height of the cone is a maximum.

20. Assuming that the work of driving a steamer through the water varies as the cube of her speed, find her most economical rate per hour against a current running c miles per hour.

Let v = speed of steamer in miles per hour.

Then av^3 = work per hour, a being constant,

and $v - c$ = actual distance advanced per hour.

Hence $\dfrac{av^3}{v - c}$ = work per mile of actual advance.

Ans. $v = \frac{3}{2}c$.

21. The sides of a triangle are a, x, y, subject to the condition $x + y = m$, a constant. Prove that the triangle of maximum area is isosceles, and that $x = y = \dfrac{m}{2}$. The area of a triangle in terms of its sides $= A = \sqrt{s(s - a)(s - x)(s - y)}$ in which $s = \frac{1}{2}$ sum of the sides. By condition, $2s = x + y + a = m + a$, whence $s = \dfrac{m + a}{2}$. Hence

$$A = \tfrac{1}{4}\sqrt{m^2 - a^2}\,\sqrt{a^2 - m^2 + 4mx - 4x^2},$$

$$A' = mx - x^2, \ f'(x) = m - 2x = 0, \ \text{and} \ x = \frac{m}{2}.$$

From $x + y = m$, $y = \dfrac{m}{2}$.

22. Find the point P of least illumination on the line joining two lights A and B, the intensity at a unit's distance of A being b, and that of B being c, knowing that the intensity varies inversely as the square of the distance.

If the distance between A and B is a, and $AP = x$, the illumination at $P = \dfrac{b}{x^2} + \dfrac{c}{(a-x)^2}$; whence $x = \dfrac{ab^{\frac{1}{3}}}{b^{\frac{1}{3}} + c^{\frac{1}{3}}}$.

23. Required the height of a light directly above the centre of a circle whose radius is R when the perimeter is most illuminated, knowing that the illumination varies directly as the sine of the angle of incidence, and inversely as the square of the distance.

Let $x = $ height.

Then the illumination $= \dfrac{x}{(x^2 - R^2)^{\frac{3}{2}}}$, $\therefore x = \dfrac{R}{\sqrt{2}}$.

24. The base of a prism is a given regular polygon whose area is A and perimeter P. The prism is surmounted by a regular pyramid whose base coincides with the head of the prism. Find the inclination θ of the faces of the pyramid to the axis, in order that the whole solid may have a given volume C with the least possible surface.

Let $x = $ height of prism. Then its volume is Ax and surface Px.

Let $a = $ perpendicular from centre of polygon on one side. Then $\dfrac{Aa\cot\theta}{3}$ is the volume of the pyramid, and $\dfrac{Pa\cosec\theta}{2}$ its surface. By condition

$$A\left(x + \frac{a\cot\theta}{3}\right) = C, \ \therefore x = \frac{C - \tfrac{1}{3}aA\cot\theta}{A}.$$

The surface which is to be a minimum is $P\left(x+\dfrac{a\cos ec\,\theta}{2}\right)$,

or, substituting the value of x, $\dfrac{C}{A}-\tfrac{1}{3}a\cot\theta+\tfrac{1}{2}a\cos ec\,\theta$.

Hence $f'(\theta)=\dfrac{a}{3}\cos ec^2\theta-\dfrac{a}{2}\cot\theta\cos ec\,\theta=0$, $\therefore 2\cos ec\,\theta=3\cot\theta$,

or $\cos\theta=\tfrac{2}{3}$, and $\theta=\cos^{-1}\tfrac{2}{3}$.

25. Prove that the minimum tangent which can be drawn to an ellipse is divided at the point of tangency into segments which are equal to the semi-axes.

If $(x,\,y)$ is the point of tangency, $\dfrac{a^2}{x}$, $\dfrac{b^2}{y}$ are the intercepts of the tangent, and the length of the tangent

$$=\sqrt{\frac{a^4}{x^2}+\frac{b^4}{y^2}}=\sqrt{\frac{a^4}{x^2}+\frac{a^2b^2}{a^2-x^2}}.$$

Hence the function to be examined is $\dfrac{a^2}{x^2}+\dfrac{b^2}{a^2-x^2}$.

From $f'(x)=0$ we find $x=a\sqrt{\dfrac{a}{a+b}}$, and from the equa-

tion of the ellipse, $y=b\sqrt{\dfrac{b}{a+b}}$. If P is the point of tan-

gency, and T the point where the tangent meets X,

$$PT=\sqrt{y^2+\left(\frac{a^2}{x}-x\right)^2}=\sqrt{\frac{b^3}{a+b}+\left(\frac{ba^{\frac{1}{2}}}{(a+b)^{\frac{1}{2}}}\right)^2}=\sqrt{\frac{b^2(a+b)}{a+b}}=b;$$

and in like manner the other segment may be shown to be a.

26. In the straight line bisecting the angle A of a triangle ABC, a point P is taken. Prove that the difference of the angles APB, APC is a maximum when AP is a mean proportional between AB and AC.

Let $AC=a$, $AB=b$, $PAB=m$, $AP=x$.

Draw PE and PF perpendicular to the sides. Then the function is

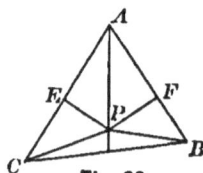

Fig. 23.

$$APC - APB = EPC - FPB = \tan^{-1}\frac{CE}{EP} - \tan^{-1}\frac{FB}{FP}$$

$$= \tan^{-1}\frac{a - x\cos m}{x\sin m} - \tan^{-1}\frac{b - x\cos m}{x\sin m};$$

whence

$$f'(x) = \frac{a}{x^2 + a^2 - 2ax\cos m} - \frac{b}{x^2 + b^2 - 2bx\cos m} = 0,$$

from which we find $x = \sqrt{ab}$.

27. A paraboloid of revolution whose axis is vertical contains a quantity of water into which is sunk a given sphere, the quantity of water being just sufficient to cover the sphere. Find the form of the paraboloid such that the quantity of water may be a minimum, knowing the volume of the paraboloid to be one-half that of the circumscribing cylinder.

Let R = radius of sphere, $z = OH$, the height of the water when the sphere is sunk, and

$$y^2 = lx \qquad (1)$$

be the equation of the parabola, in which l is the unknown parameter. The equation of the circle is

$$y^2 + (x - OC)^2 = R^2,$$

or, since $OC = z - R$,

$$y^2 + (x - z + R)^2 = R^2. \qquad (2)$$

Combining (1) and (2), we have

$$lx + (x - z + R)^2 = R^2,$$

whence $\quad x = \dfrac{2(z - R) - l \pm \sqrt{4R^2 + l^2 - 4zl + 4Rl}}{2}.$

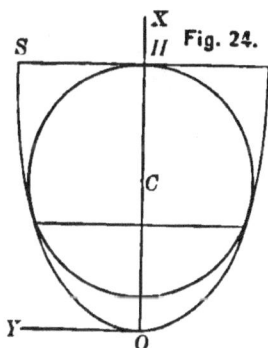

But the circle is tangent to the parabola, and there can be but one value for x, and hence

$$4 R^2 + l^2 - 4 zl + 4 Rl = 0,$$

$$\therefore z = \frac{(l + 2 R)^2}{4 l}. \tag{3}$$

The vol. of water = vol. of paraboloid — vol. of sphere

$$= \frac{\pi}{2} HS^2 \cdot HO - \tfrac{4}{3} \pi R^3 = \tfrac{1}{2} \pi l z^2 - \tfrac{4}{3} \pi R^3 = V,$$

since (1) gives $HS^2 = lz$. Substituting the value of z from (3), the function becomes

$$V = \frac{\pi}{2} \frac{(l + 2 R)^4}{16 l} - \tfrac{4}{3} \pi R^3,$$

or $$V' = \frac{(l + 2 R)^4}{l};$$

whence $f'(l) = 0$ gives $l = \tfrac{2}{3} R$, which determines the form of the paraboloid.

CHAPTER IV.

FUNCTIONS OF TWO OR MORE VARIABLES.

87. A partial differential *of a function of two or more variables is its differential on the hypothesis that only one of the variables changes.* Thus, if $u = \sin x \log y + zx^2$, and x only is supposed to change, y and z being regarded constant, the partial differential of u with respect to x is $\cos x \log y dx + 2zx dx$; and the partial differentials of u with respect to y and z are $\dfrac{\sin x}{y} dy$ and $x^2 dz$, respectively.

88. Notation. To distinguish the partial differentials, the variable with respect to which the function is differentiated is written as a subscript, thus :

$$d_x u = (\cos x \log y + 2zx)dx, \quad d_y u = \frac{\sin x}{y} dy, \quad d_z u = x^2 dz,$$

which are read 'the x-differential of u,' etc.

$\dfrac{d_x u}{dt}$ is evidently the rate of the function so far as its rate depends upon the rate of x.

89. A partial derivative, *or* partial differential coefficient, *is the ratio of a partial differential to the differential of the variable which is supposed to vary.* Thus, in the above case, the partial derivatives of u with respect to x, y, and z, respectively, are

$$\frac{du}{dx} = \cos x \log y + 2zx, \quad \frac{du}{dy} = \frac{\sin x}{y}, \quad \frac{du}{dz} = x^2,$$

the subscripts being omitted as the denominators indicate the variable with respect to which the differentiation is performed.

111

A partial derivative is the ratio of the rate of the function to that of the variable supposed to vary so far as the rate of the former depends upon that of the latter, and the function is an increasing or a decreasing function of any one of its variables according as the corresponding partial derivative is positive or negative (Art. 22).

Since the first derivative is the factor by which the differential of the variable is multiplied to obtain the differential of the function, the partial differentials are also represented by the notation

$$\frac{du}{dx}\,dx,\quad \frac{du}{dy}\,dy,\quad \frac{du}{dz}\,dz,\ \text{etc.,}$$

which are equivalent to $d_z u$, $d_y u$, $d_z u$, etc.

EXAMPLES.   Find the partial differentials of :

1. $u = (x^2 + y^2)^{\frac{1}{2}}$. $\dfrac{du}{dx}\,dx = \dfrac{x\,dx}{(x^2 + y^2)^{\frac{1}{2}}},\quad \dfrac{du}{dy}\,dy = \dfrac{y\,dy}{(x^2 + y^2)^{\frac{1}{2}}}.$

2. $u = \sin^{-1}\dfrac{x}{y}$. $\dfrac{du}{dx}\,dx = \dfrac{dx}{\sqrt{y^2 - x^2}},\quad \dfrac{du}{dy}\,dy = - \dfrac{x\,dy}{y\sqrt{y^2 - x^2}}.$

3. $u = y^{xz}$. $\dfrac{du}{dx}\,dx = zy^{xz} \log y\,dx,\quad \dfrac{du}{dy}\,dy = xzy^{xz-1}\,dy,$

$$\frac{du}{dz}\,dz = xy^{xz} \log y\,dz.$$

Find the partial derivatives of :

4. $u = \sin(xy)$. $\dfrac{du}{dx} = y \cos(xy),\quad \dfrac{du}{dy} = x \cos(xy).$

5. $u = y^{\sin x}$. $\dfrac{du}{dx} = y^{\sin x} \log y \cos x,$

$$\frac{du}{dy} = \sin x \cdot y^{\sin x-1} = \frac{\sin x}{y^{\text{covers} z}}.$$

6. $u = \log(x + y)$. $\dfrac{du}{dx} = \dfrac{du}{dy} = \dfrac{1}{x + y}.$

90. The total differential *of a function is its differential obtained on the hypothesis that all its variables change.*

Since the total differential of $u = f(x, y, z,$ etc.) can contain only the first powers of dx, dy, dz, etc., it will be of the form $du = f_1(x, y, z,$ etc.$)dx + f_2(x, y, z,$ etc.$)dy + f_3(x, y, z,$ etc.$)dz$ + etc., in which $f_1(x, y, z,$ etc.$)$, $f_2(x, y, z,$ etc.$)$, etc., represent the collected coefficients of dx, dy, etc. But if all the variables except x be regarded constant, $dy = dz =$ etc. $= 0$, and all the terms vanish except the first, which is the partial differential with respect to x; and if all the variables except y be regarded constant, all the terms vanish except the second, which is the partial differential with respect to y; and so on. Thus all the terms of the second member will be obtained by differentiating u in succession as if all the variables but one were constant. Hence *the total differential of a function is the sum of its partial differentials.*

ILLUSTRATION. Let $u = 3ax^2y + z^3e^z$.

$$du = d_xu + d_yu + d_zu$$
$$= (6axy + z^3e^z)dx + 3ax^2dy + 3z^2e^zdz. \qquad (1)$$

Equation (1) is evidently true whether the variables be dependent or independent. If, however, the variables are dependent, the total differential may be expressed in terms of any one of them. Thus if

$$y = bx,\ z = \sin x,\ \text{whence}\ dy = bdx,\ dz = \cos x dx,$$

(1) becomes

$$du = (9abx^2 + e^z \sin^3 x - 3e^z \sin^2 x \cos x)dx;$$

or the same result might be obtained by substituting the values of y and z in the original function, giving

$$u = 3abx^3 + e^z \sin^3 x,$$

whence, differentiating, we have (2), as before.

If Equation (1) be divided by dt, we have the rate of the

function in terms of the rates of its variables; and if the variables are dependent, the rate of the function may be expressed in terms of that of any one of its variables assumed as the independent variable. Thus, dividing (2) by dt, we have the rate of u in terms of the rate of x.

EXAMPLES.   Find the total differentials of:

1. $u = ax^2y + by^3x$. $du(3ax^2y + by^3)dx + (ax^3 + 3by^2x)dy$.

2. $u = \tan^{-1}\dfrac{x}{y}$. $du = \dfrac{ydx - xdy}{x^2 + y^2}$.

3. $u = \log x^y$. $du = \dfrac{ydx}{x} + \log xdy$.

4. $u = r\cos\theta$. $du = \cos\theta dr - r\sin\theta dr$.

5. $u = \dfrac{xy}{x + y}$. $du = \dfrac{x^2dy + y^2dx}{(x + y)^2}$.

6. $u = e^x a^y$. $du = a^y e^x dx + e^x a^y \log ady$.

7. $u = \tan^{-1}(xy)$. $du = \dfrac{ydx + xdy}{1 + x^2y^2}$.

8. $u = x^y$. $du = yx^{y-1}dx + x^y \log xdy$.

9. $u = a^x + e^{-y}z + \sin v$.

$du = a^x \log adx - ze^{-y}dy + e^{-y}dz + \cos vdv$.

10. $u = \tan^{-1}\dfrac{x - y}{x + y}$. $du = \dfrac{ydx - xdy}{x^2 + y^2}$.

91. The total derivative, *or* **total differential coefficient,** *of a function is the ratio of the total differential to the differential of the independent variable.*

Thus, if $u = f(x, y, z)$, we have for the total differential, by Art. 90,

$$du = \frac{du}{dx}dx + \frac{du}{dy}dy + \frac{du}{dz}dz; \qquad (1)$$

and if x be the independent variable,

$$\left[\frac{du}{dx}\right] = \frac{du}{dx} + \frac{du}{dy}\frac{dy}{dx} + \frac{du}{dz}\frac{dz}{dx}. \qquad (2)$$

The student will observe that the du's are not the same in the above formulæ. In the first member of (1) du is the total differential of u, while in the second member du is a partial differential, the notation serving to distinguish the partial differentials from each other and from the total differential.

To cancel the equal factors from the terms of the second member would be to destroy the means of distinguishing the du's, no two of which are the same. In fact, (1) is

$$du = d_x u + d_y u + d_z u.$$

In forming (2) from (1) the first member becomes a total derivative, the bracket being used to distinguish it from the partial derivative $\dfrac{du}{dx}$ in the second member. It is further to be observed that while (1) is true whether the variables be dependent or independent, (2) has no significance unless the variables are dependent; for $\dfrac{dy}{dx}, \dfrac{dz}{dx}$, cannot be evaluated unless $y = \phi(x), z = \psi(x)$.

The total derivative is evidently the ratio of the rate of the function, on the hypothesis that all its variables change to the rate of the independent variable.

92. The total derivative with respect to any independent variable may be formed in like manner by dividing the total differential, which is always the sum of the partial differentials, by the differential of the independent variable, understanding that the independent variable is connected with the others by auxiliary relations.

Thus, given $u = f(x, y, z)$ and $x = \phi_1(w), y = \phi_2(w), z = \phi_3(w)$, to form the total derivative with respect to w, we have

$$du = \frac{du}{dx}\,dx + \frac{du}{dy}\,dy + \frac{du}{dz}\,dz,$$

whence
$$\left[\frac{du}{dw}\right] = \frac{du}{dx}\frac{dx}{dw} + \frac{du}{dy}\frac{dy}{dw} + \frac{du}{dz}\frac{dz}{dw}.$$

In this case the bracket is not necessary, but it is usual to enclose the total derivative in brackets whatever the independent variable.

EXAMPLES. 1. Given $u = 2axy + \log x$, and $x = \sin y$, to form the total derivative of u with respect to y.

$$du = \frac{du}{dx}dx + \frac{du}{dy}dy,$$

$$\left[\frac{du}{dy}\right] = \frac{du}{dx}\frac{dx}{dy} + \frac{du}{dy}.$$

From the given function we find the partial derivatives

$$\frac{du}{dx} = 2ay + \frac{1}{x}, \quad \frac{du}{dy} = 2ax,$$

and from $x = \sin y$, $\dfrac{dx}{dy} = \cos y$. Substituting these values,

$$\left[\frac{du}{dy}\right] = \left(2ay + \frac{1}{x}\right)\cos y + 2ax$$
$$= 2a(y\cos y + \sin y) + \cot y.$$

The same result would be obtained by first substituting the value of x in the function and then differentiating.

2. $u = y^2 + z^4 + zy$, $y = \sin x$, $z = \cos x$.

$$\left[\frac{du}{dx}\right] = \cos 2x(1 - \sin 2x).$$

3. $u = \tan^{-1}\dfrac{x-y}{x+y}$, $x = e^z$, $y = e^{-z}$.

$$\left[\frac{du}{dz}\right] = \frac{2e^{2z}}{e^{4z}+1}.$$

By substituting the values of x and y in u and then differentiating, the student may compare the two processes. In this case the use of the formula is more expeditious.

4. $u = \tan^{-1}\dfrac{x}{y}$, $x^2 + y^2 = R^2$.

$$\left[\frac{du}{dy}\right] = -\frac{1}{x}.$$

5. $u = \sin\dfrac{z}{y}$, $z = e^x$, $y = x^2$.

$$\left[\frac{du}{dx}\right] = (x-2)\frac{e^x}{x^3}\cos\frac{e^x}{x^2}.$$

6. $u = yz$, $y = e^x$, $z = x^4 - 4x^3 + 12x^2 - 24x + 24$.

$$\left[\frac{du}{dx}\right] = e^x x^4.$$

93. Implicit function of two variables.

Let $f(x, y) = 0$. Representing the function by u we have

$$u = f(x, y) = 0.$$

Since the only possible values of the variables are those which render the function zero, u is constant, and hence its differential is zero. Therefore

$$du = \frac{du}{dx}dx + \frac{du}{dy}dy = 0,$$

whence

$$\frac{dy}{dx} = -\frac{\dfrac{du}{dx}}{\dfrac{du}{dy}}, \tag{1}$$

or *the first derivative of an implicit function is the negative ratio of its partial derivatives.*

The above depends solely upon the fact that du is constant; hence (1) is true when $f(x, y) = a$, where a is any constant. Thus, $y^2 - 2px = 0$, $\therefore \dfrac{dy}{dx} = -\dfrac{-2p}{2y} = \dfrac{p}{y}$. Again, $y^2 + x^2 = R^2$, $\therefore \dfrac{dy}{dx} = -\dfrac{2x}{2y} = -\dfrac{x}{y}$. These results might of course be obtained directly by the ordinary process of differentiation, but it is often useful to employ the value of the derivative in terms of the partial derivatives as given in (1).

EXAMPLES.   Form by the above method the derivatives of:

1. $u = 3\,ax^2y - 2\,ay^2x = c.$

$$\frac{dy}{dx} = -\frac{\dfrac{du}{dx}}{\dfrac{du}{dy}} = -\frac{6\,axy - 2\,ay^2}{3\,ax^2 - 4\,ayx} = \frac{y(2\,y - 6\,x)}{x(3\,x - 4\,y)}.$$

2. $u = x\log y - y\log x = 0.$ $\dfrac{dy}{dx} = \dfrac{yx\log y - y^2}{yx\log x - x^2}.$

3. $u = y^3 + x^3 - 3\,mxy = 0.$ $\dfrac{dy}{dx} = \dfrac{my - x^2}{y^2 - mx}.$

4. $u = ye^{ny} - ax^m = 0.$ $\dfrac{dy}{dx} = \dfrac{my}{x(1 + ny)}.$

5. $u = \sin(xy) + \tan(xy) = a.$ $\dfrac{dy}{dx} = -\dfrac{y}{x}.$

6. $u = ay^3 - x^3y - ax^3 = 0.$ $\dfrac{dy}{dx} = \dfrac{3\,x^2y + 3\,ax^2}{3\,ay^2 - x^3}.$

94. Evaluation of the first derivative of an implicit function.

Let $f(x, y) = 0$, in which x is the equicrescent variable.

Then
$$\frac{dy}{dx} = -\frac{\dfrac{du}{dx}}{\dfrac{du}{dy}}$$

may be a function of both x and y and assume the illusory form $\dfrac{0}{0}$ for particular values of x and y.   In such a case we may eliminate y from $\dfrac{dy}{dx}$ by means of $f(x, y) = 0$, and then proceed as in Art. 74; or, since y is a function of x, we may apply the process of Art. 74 directly, without eliminating y, forming the successive derivatives of the numerator and denominator with respect to x until a pair is found whose ratio does not become $\dfrac{0}{0}$ for the particular values of the variables.   Thus, if $u = y^3 - x^2y - x^4 = 0$,

$$\frac{dy}{dx} = -\frac{\dfrac{du}{dx}}{\dfrac{du}{dy}} = \frac{2\,xy + 4\,x^3}{3\,y^2 - x^2},$$

which for $x = y = 0$ becomes $\dfrac{0}{0}$. By Art. 74,

$$\frac{dy}{dx} = \frac{2\,xy + 4\,x^3}{3\,y^2 - x^2}\Bigg]_{0,\,0} = \frac{2\,y + 2\,x\dfrac{dy}{dx} + 12\,x^2}{6\,y\dfrac{dy}{dx} - 2\,x}\Bigg]_{0,\,0}$$

$$= \frac{2\dfrac{dy}{dx} + 2\,x\dfrac{d^2y}{dx^2} + 2\dfrac{dy}{dx} + 24\,x}{6\dfrac{dy^2}{dx^2} + 6\,y\dfrac{d^2y}{dx^2} - 2}\Bigg]_{0,\,0} = \frac{4\dfrac{dy}{dx}}{6\dfrac{dy^2}{dx^2} - 2}.$$

Hence $6\dfrac{dy^3}{dx^3} - 2\dfrac{dy}{dx} = 4\dfrac{dy}{dx}$, or $\dfrac{dy}{dx} = 0$, and ± 1.

Having seen that the true value of an expression which assumes the form $\dfrac{0}{0}$ for a particular value of the variable is its limit (Art. 74), it would seem, since a quantity can have but one limit (Art. 64), that $\dfrac{dy}{dx}$ in the above example could have but one value. That it may have several values will appear in Art. 114.

EXAMPLES. 1. If $u = x^4 + 2\,ax^2y - ay^3 = 0$, show that $\dfrac{dy}{dx} = 0$, or $\pm\sqrt{2}$ when $x = y = 0$.

$$\frac{dy}{dx} = -\frac{4\,x^3 + 4\,axy}{2\,ax^2 - 3\,ay^2}\Bigg]_{0,\,0} = \frac{12\,x^2 + 4\,ay + 4\,ax\dfrac{dy}{dx}}{6\,ay\dfrac{dy}{dx} - 4\,ax}\Bigg]_{0,\,0}$$

$$= \frac{24\,x + 4\,a\dfrac{dy}{dx} + 4\,a\dfrac{dy}{dx} + 4\,ax\dfrac{d^2y}{dx^2}}{6\,a\dfrac{dy^2}{dx^2} + 6\,ay\dfrac{d^2y}{dx^2} - 4\,a}\Bigg]_{0,\,0} = \frac{8\,a\dfrac{dy}{dx}}{6\,a\dfrac{dy^2}{dx^2} - 4\,a},$$

whence $\dfrac{dy}{dx}\left(\dfrac{dy^2}{dx^2} - 2\right) = 0$.

2. If $u = x^4 - ay^3 + 2axy^2 + 3ax^2y = 0$, show that $\dfrac{dy}{dx} = 0, 3$, or -1, when $x = y = 0$.

3. $u = x^4 - a^2xy + y^2 = 0$.   Prove that $\dfrac{dy}{dx} = 0$, or a^2, when $x = y = 0$.

4. $u = x^4 + ax^2y - ay^3 = 0$.   Prove that $\dfrac{dy}{dx} = 0$, or ± 1, when $x = y = 0$.

5. $u = a^3y^2 - 2abx^2y - x^5 = 0$.   Prove that $\dfrac{dy}{dx} = \pm 0$, when $x = y = 0$.

CHAPTER V.

CURVATURE.

95. *A curve is concave upward at any of its points when its tangent at that point lies below the curve, and is convex upward when its tangent lies above the curve.*

When a curve is concave upward, its slope increases with x; hence if $y = f(x)$ be its equation, $f'(x) + \dfrac{dy}{dx}$ is an increasing function. But the first derivative of an increasing function is positive, and the first derivative of $f'(x)$ is $f''(x) = \dfrac{d^2y}{dx^2}$. Hence $f''(x)$ *is positive when the curve is concave upward.*

If the curve is convex upward, its slope decreases as x increases ; $f'(x)$ is a decreasing function, and its derivative is therefore negative. Hence $f''(x)$ *is negative when the curve is convex upward.*

96. A point at which, as x increases, the curvature changes from concave to convex upward, or *vice versa*, is called a **point of inflexion.** At a point of inflexion the tangent evidently cuts the curve.

Since on one side the curve is convex and on the other concave upward, *the analytic condition for a point of inflexion is*

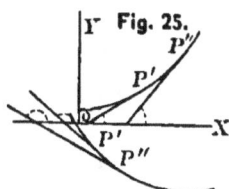

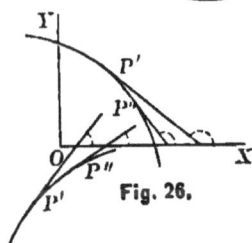

121

(Art. 95) *a change of sign in* $f''(x)$. Hence all values of x corresponding to such points are roots of the equations $f''(x) = 0$, $f''(x) = \infty$. These roots are critical values, and do not correspond to points of inflexion unless accompanied by a change of sign in $f''(x)$.

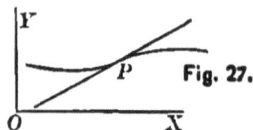

In approaching a point of inflexion $f'(x)$ is increasing (or decreasing), and after passing this point is decreasing (or increasing); hence $f'(x)$ is either a maximum or a minimum at a point of inflexion.

EXAMPLES. Examine the following curves for curvature and points of inflexion:

1. $x = \log y$, the logarithmic curve.

$f''(x) = y$, which is always positive, since negative numbers have no logarithms. The curve is therefore always concave upward and has no point of inflexion.

2. $y^2 + x^2 = R^2$, the circle.

$f''(x) = -\dfrac{R^2}{y^3}$, which is negative when y is positive, and positive when y is negative; hence the curve is convex upward above, and concave upward below, X. $f''(x)$ has two signs, but does not change sign for increasing values of x, and there is no point of inflexion.

3. $xy = m$, the hyperbola.

$f''(x) = \dfrac{2m}{x^3}$, which has the sign of x. The curve is therefore concave upward in the first, and convex upward in the third, angle. $f''(x)$ changes sign at $x = 0$; but when $x = 0$, $y = \infty$, the curve being discontinuous, and there is no point of inflexion.

4. $x^2 y = 4a^2(2a - y)$, the witch.

$f''(x) = 2y\dfrac{3x^2 - 4a^2}{(x^2 + 4a^2)^2}$. Points of inflexion at $x = \pm\dfrac{2a}{\sqrt{3}}$.

5. $ay^2 = x^3$, the semi-cubical parabola.

$$f''(x) = \frac{3x^4}{4 a^2 y^3}.$$

6. $y = \sin x$, the sinusoid.

7. $x = \log^3 y$. A point of inflexion at $x = 8$, where the curvature changes from convex to concave upward.

8. $y^3 = a^2 x$, the cubical parabola.

$f''(x) = -\frac{2 a^4}{9 y^5}$, which is positive when y is negative, and negative when y is positive; hence the curve is concave upward in the third, and convex upward in the first, angle. $f''(x)$ changes sign at $y = 0$, whence $x = 0$, passing through infinity, and the origin is a point of inflexion.

9. $y(a^4 - b^4) = x(x - a)^4 - x b^4$.
A point of inflexion when $x = \frac{2}{3} a$.

10. $y = \frac{x^3}{a^2 + x^2}.$

11. $y = \tan x$.

97. Rate of curvature. A plane curve may be defined as the locus of a point which always moves along a straight line while the line always turns around the point.

Since the direction of motion is always that of the line, the line is the tangent to the curve. Were the line to remain fixed, the locus would be a straight line, that is, if the tangent does not turn about the moving point there is no curvature; hence, if ϕ be the angle which the tangent makes with any fixed line as X, the curvature will depend upon the change of ϕ.

Since in the circle equal arcs subtend equal angles at the centre, the normal, and therefore the tangent, turns through the same angle for every unit of path described by the generating point, and the curvature of the circle is therefore constant whatever the unit by which it is measured.

It is evident that if, in passing a second time through any point of a given curve, the velocity of the generating point be m times what it was before, the rate of turning of the tangent at that point will also be m times its former rate; or that the ratio of the rate of turning of the tangent to the velocity in the curve is constant.   Hence

$$\frac{\dfrac{d\phi}{dt}}{\dfrac{ds}{dt}} = \frac{d\phi}{ds}$$

is a constant for the same point, whatever the velocity.   This expression is evidently the rate of turning of the tangent *per unit of length of the curve*, and may be taken as a measure of the curvature.   This measure is independent of t, as it should be, for the curvature is a geometric property of the curve independent of the time of its description.

Since the rate $\dfrac{d\phi}{ds}$ is the amount by which ϕ would change for a unit's length of path, were its rate to remain through this distance what it was at its beginning, the curvature at any point of a plane curve is that of a circle which has a common tangent with the curve at the point considered.   This circle is called the **circle of curvature**, and its radius the **radius of curvature**.

98.   *To express* $\dfrac{d\phi}{ds}$ *in terms of the coordinates of the generating point.*

From Art. 25 we have

$$ds = \sqrt{dx^2 + dy^2},$$

and, reckoning ϕ from the axis of X,

$$\tan\phi = \frac{dy}{dx}, \text{ whence } \sec^2\phi\, d\phi = \frac{d^2\phi}{dx},$$

or

$$d\phi = \frac{\dfrac{d^2y}{dx}}{\sec^2\phi} = \frac{\dfrac{d^2y}{dx}}{1 + \tan^2\phi} = \frac{\dfrac{d^2y}{dx}}{1 + \dfrac{dy^2}{dx^2}}.$$

Hence $$\frac{d\phi}{ds} = \frac{\frac{d^2y}{dx}}{\left(1 + \frac{dy^2}{dx^2}\right)(dx^2 + dy^2)^{\frac{1}{2}}} = \frac{\frac{d^2y}{dx^2}}{\left(1 + \frac{dy^2}{dx^2}\right)^{\frac{3}{2}}}. \qquad (1)$$

To find therefore the curvature of a plane curve $y = f(x)$, differentiate its equation twice and substitute in (1) the values of the first and second derivatives.

99. Curvature of the circle.

From $x^2 + y^2 = R^2, \dfrac{dy}{dx} = -\dfrac{x}{y} = -\dfrac{x}{\sqrt{R^2 - x^2}}; \dfrac{d^2y}{dx^2} = -\dfrac{R}{y^3}$, which will be $\pm$ as y is $\mp$. Hence

$$\frac{d\phi}{ds} = \frac{\frac{R^2}{y^3}}{\left(1 + \frac{x^2}{y^2}\right)^{\frac{3}{2}}} = \pm \frac{1}{R}.$$

or the curvature of a circle is the reciprocal of its radius.

Cor. 1. Since $\dfrac{d\phi}{ds} = 1$ when $R = 1$, the *unit* of curvature is seen to be the curvature of the circle whose radius is unity.

Cor. 2. The curvatures of any two circles are inversely as their radii.

100. Radius of curvature.

Since the curvature of any plane curve at a given point is that of its circle of curvature at that point, and the curvature of this circle is measured by the reciprocal of its radius, we have, if ρ be this radius,

$$\frac{1}{\rho} = \frac{d\phi}{ds}.$$

or $$\rho = \frac{ds}{d\phi} = \frac{\left(1 + \frac{dy^2}{dx^2}\right)^{\frac{3}{2}}}{\frac{d^2y}{dx^2}}.$$

If we take the positive value of the radical, the radius of curvature will be $\pm$ as $\dfrac{d^2y}{dx^2}$ is $\pm$; that is, according as the curve is concave upward or downward at the point considered. The sign of ρ may thus serve to determine the *direction* of curvature.

Cor. 1. Since $\dfrac{d^2y}{dx^2} = 0$ at a point of inflexion, the radius of curvature at a point of inflexion is infinite, and the curvature zero.

Cor. 2. Since the circle of curvature at any point has a tangent in common in the curve, the radius of curvature is a normal to the curve.

101. Coordinates of the centre of curvature.

Let C be the centre of curvature of the curve MN at any point P, and a, β the co-ordinates of C. Then

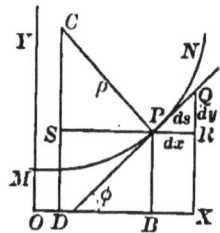

Fig. 28.

$$a = OD = OB - DB = x - \rho \sin \phi$$
$$= x - \rho \frac{dy}{ds},$$
$$\beta = DC = BP + SC = y + \rho \cos \phi$$
$$= y + \rho \frac{dx}{ds}.$$

Substituting the values of

$$ds = \sqrt{dx^2 + dy^2} \quad \text{and} \quad \rho = \frac{\left[1 + \left(\dfrac{dy}{dx}\right)^2\right]^{\frac{3}{2}}}{\dfrac{d^2y}{dx^2}},$$

$$a = x - \frac{\left[1 + \left(\dfrac{dy}{dx}\right)^2\right]\dfrac{dy}{dx}}{\dfrac{d^2y}{dx^2}}, \quad \beta = y + \frac{1 + \left(\dfrac{dy}{dx}\right)^2}{\dfrac{d^2y}{dx^2}}. \tag{1}$$

102. Maximum or minimum curvature. Since the curvature is measured by $\dfrac{1}{\rho}$, it will be a maximum or minimum when ρ is a minimum or maximum. It is further evident that if a curve is symmetrical with reference to the normal in the vicinity of the point of contact, the curvature, if not constant, will be a maximum or a minimum at that point.

103. *In the vicinity of a point of maximum or minimum curvature, the circle of curvature lies wholly on one side of the curve; at all other points it intersects the curve.* For at a point of maximum curvature the rate of turning of the tangent is greater than immediately before or after, while the rate of turning of the tangent to the circle of curvature remains constantly what it was at the point of contact; hence the circle lies within the curve at this point. For a like reason, at a point of minimum curvature, the circle lies without the curve. At all other points of the locus (except when it is a straight line) its curvature is continually increasing (or decreasing) while that of the circle remains the same; on one side, therefore, the curvature is less and on the other greater than that of the circle, and hence the curve crosses the circle.

Thus, the circle of curvature lies without the ellipse at the extremities of the conjugate axis, within at the extremities of the transverse axis, and at all other points cuts the ellipse.

EXAMPLES. 1. The parabola.

From $y^2 = 2px$, $\dfrac{dy}{dx} = \dfrac{p}{y}$, $\dfrac{d^2y}{dx^2} = -\dfrac{p^2}{y^3}$.

Hence
$$\rho = \frac{\left[1 + \left(\dfrac{dy}{dx}\right)^2\right]^{\frac{3}{2}}}{\dfrac{d^2y}{dx^2}} = \frac{(y^2 + p^2)^{\frac{3}{2}}}{p^2},$$

$$a = x - \frac{\left[1 + \left(\dfrac{dy}{dx}\right)^2\right]\dfrac{dy}{dx}}{\dfrac{d^2y}{dx^2}} = 3x + p,$$

$$\beta = y + \frac{1 + \left(\frac{dy}{dx}\right)^2}{\frac{d^2y}{dx^2}} = -\frac{y^3}{p^2}.$$

At the vertex $y \doteq x = 0$, $\rho = p$, $a = p$, $\beta = 0$, or the radius of curvature is one-half the parameter and the centre of curvature on the axis twice as far from the vertex as the focus is. We observe also that ρ is least when $y = 0$, or the curvature at the vertex is a maximum.

2. The ellipse.

From $a^2y^2 + b^2x^2 = a^2b^2$, $\dfrac{dy}{dx} = -\dfrac{b^2x}{a^2y}$, $\dfrac{d^2y}{dx^2} = -\dfrac{b^4}{a^2y^3}$.

Hence $\rho = \dfrac{(a^4y^2 + b^4x^2)^{\frac{3}{2}}}{a^4b^4}$.

At the extremities of the conjugate axis, $x = 0$, $y = \pm b$,
$\rho = \dfrac{a^2}{b}$.

At the extremities of the transverse axis, $y = 0$, $x = \pm a$,
$\rho = \dfrac{b^2}{a}$.

If $a = b = R$, $\rho = R$, the radius of the circle.

3. The cycloid.

From $x = r \operatorname{vers}^{-1}\dfrac{y}{r} - \sqrt{2ry - y^2}$, $\dfrac{dy}{dx} = \dfrac{\sqrt{2ry - y^2}}{y}$, $\dfrac{d^2y}{dx^2} = -\dfrac{r}{y^2}$.

Hence $\rho = 2\sqrt{2ry}$, or the radius of curvature is twice the corresponding normal (Art. 48, Ex. 9).

At the highest point, $y = 2r$, $\rho = 4r$; at the vertex, $y = 0$, $\rho = 0$.

4. The logarithmic curve.

From $x = \log_a y$, $\dfrac{dy}{dx} = \dfrac{y}{m}$, $\dfrac{d^2y}{dx^2} = \dfrac{y}{m^2}$; hence $\rho = \dfrac{(m^2 + y^2)^{\frac{3}{2}}}{my}$.

If $a = e$, $\rho = \dfrac{(1 + y^2)^{\frac{3}{2}}}{y}$, and if $x = 0$, whence $y = 1$, $\rho = 2\sqrt{2}$, the radius of curvature of the Naperian curve at the point where it crosses Y.

5. The hypocycloid.

From $x^{\frac{2}{3}} + y^{\frac{2}{3}} = a^{\frac{2}{3}}$, $\dfrac{dy}{dx} = -\dfrac{y^{\frac{1}{3}}}{x^{\frac{1}{3}}}$, $\dfrac{d^2y}{dx^2} = \dfrac{1}{3}\dfrac{a^{\frac{2}{3}}}{y^{\frac{1}{3}}x^{\frac{4}{3}}}$; hence $\rho = 3\sqrt[3]{axy}$.

When either x or y is zero, $\rho = 0$.

6. The cubical parabola.

From $y^3 = a^2x$, $\rho = \dfrac{(9y^4 + a^4)^{\frac{3}{2}}}{6a^4y}$.

7. The semi-cubical parabola.

From $ay^2 = x^3$, $\rho = \dfrac{(4a + 9x)^{\frac{3}{2}}}{6a}x^{\frac{1}{2}}$.

8. The catenary.

From $y = \dfrac{a}{2}(e^{\frac{x}{a}} + e^{-\frac{x}{a}})$, $\rho = -\dfrac{y^2}{a}$.

9. The cissoid.

From $y^2 = \dfrac{x^3}{2a - x}$, $\rho = \dfrac{a(8a - 3x)^{\frac{3}{2}}x^{\frac{1}{2}}}{3(2a - x)^2}$, which is zero when $x = 0$, and infinity when $x = 2a$.

EVOLUTES AND ENVELOPES.

104. *The locus of the centre of curvature of a given curve is called the* **evolute** *of the curve.*

The given curve is called the **involute.**

105. Equation of the evolute.

Let $y = f(x)$ be the equation of the involute. The coordinates of its centre of curvature are (Art. 101),

$$a = x - \frac{\left[1 + \left(\dfrac{dy}{dx}\right)^2\right]\dfrac{dy}{dx}}{\dfrac{d^2y}{dx^2}}, \quad \beta = y + \frac{1 + \left(\dfrac{dy}{dx}\right)^2}{\dfrac{d^2y}{dx^2}}.$$

By substituting in these the values of the derivatives obtained from $y = f(x)$, we obtain the values of a and β in terms of x and y. Eliminating x and y between these results and $y = f(x)$, the resulting equation between a and β will be the equation of the evolute.

EXAMPLES. 1. Find the equation of the evolute of the parabola.

From Art. 103, Ex. 1, we have

$$a = 3x + p, \quad \beta = -\frac{y^3}{p^2},$$

whence

$$x = \frac{a - p}{3}, \quad y = -\beta^{\frac{1}{3}} p^{\frac{2}{3}}.$$

Substituting these values of x and y in $y^2 = 2px$, we have

$$\beta^2 = \frac{8}{27 p} (a - p)^3.$$

The form of the evolute is shown in the figure.

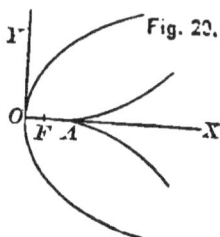

2. Find the equation of the evolute to the ellipse.

$$a = \frac{(a^2 - b^2) x^3}{a^4}, \quad \beta = -\frac{(a^2 - b^2) y^3}{b^4},$$

whence $x = \left(\dfrac{a^4 a}{a^2 - b^2}\right)^{\frac{1}{3}}, \quad y = -\left(\dfrac{b^4 \beta}{a^2 - b^2}\right)^{\frac{1}{3}}.$

Substituting these in $a^2 y^2 + b^2 x^2 = a^2 b^2$, we find

$$(aa)^{\frac{2}{3}} + (b\beta)^{\frac{2}{3}} = (a^2 - b^2)^{\frac{2}{3}},$$

and the form of the curve is shown in the figure.

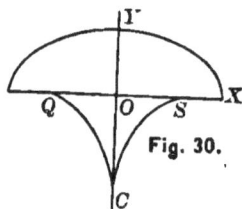

3. The evolute of the cycloid is an equal cycloid.

From $x = r \text{ vers}^{-1} \dfrac{y}{r} - \sqrt{2 ry - y^2}$, $\dfrac{dy}{dx} = \dfrac{\sqrt{2 ry - y^2}}{y}$, $\dfrac{d^2 y}{dx^2} = -\dfrac{r}{y^2}$.

Hence $x = a - 2\sqrt{-2r\beta - \beta^2}$, $y = -\beta$.

Substituting these in the equation of the cycloid,

$$a = r \text{ vers}^{-1}\left(-\frac{\beta}{r}\right) + \sqrt{-2 r\beta - \beta^2}. \tag{1}$$

If the given cycloid be referred to the axes $X_1 O_1 Y_1$.

$$O_1 N = x = CD + QP = MP + QP = MP + \sqrt{MQ \cdot QD}$$
$$= r \operatorname{vers}^{-1}\left(\frac{-y}{r}\right) + \sqrt{-2ry - y^2},$$

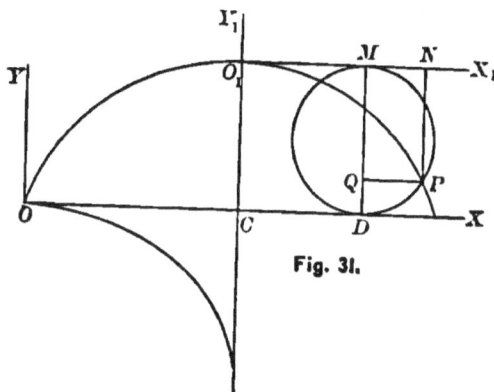

Fig. 31.

which is of the same form as (1). Hence the evolute is an equal cycloid, O being its highest point.

4. Show that the evolute of a circle is a point, the centre of the circle.

The usefulness of the above method of finding the equation of the evolute is limited by the difficulties of elimination, although the method is general.

5. To find the evolute of the hypocycloid.

From $x^{\frac{2}{3}} + y^{\frac{2}{3}} = a^{\frac{2}{3}}$, we have $\dfrac{dy}{dx} = -\dfrac{y^{\frac{1}{3}}}{x^{\frac{1}{3}}}$, $\dfrac{d^2 y}{dx^2} = \dfrac{1}{3}\dfrac{a^{\frac{2}{3}}}{y^{\frac{1}{3}}x^{\frac{4}{3}}}$.

Hence $a = x + 3x^{\frac{1}{3}}y^{\frac{2}{3}}$, $\beta = y + 3y^{\frac{1}{3}}x^{\frac{2}{3}}$.

To eliminate x and y we proceed as follows:

$$a + \beta = x + 3x^{\frac{1}{3}}y^{\frac{2}{3}} + 3y^{\frac{1}{3}}x^{\frac{2}{3}} + y = (x^{\frac{1}{3}} + y^{\frac{1}{3}})^3;$$

hence $\qquad (a + \beta)^{\frac{1}{3}} = x^{\frac{1}{3}} + y^{\frac{1}{3}}$.

Similarly, $\quad (a - \beta)^{\frac{1}{3}} = x^{\frac{1}{3}} - y^{\frac{1}{3}}$.

Hence, $\qquad (a + \beta)^{\frac{1}{3}} + (a - \beta)^{\frac{1}{3}} = 2x^{\frac{1}{3}}$,

$$(\alpha + \beta)^{\frac{1}{3}} - (\alpha - \beta)^{\frac{1}{3}} = 2\,y^{\frac{1}{3}},$$

and
$$[(\alpha + \beta)^{\frac{1}{3}} + (\alpha - \beta)^{\frac{1}{3}}]^2 + [(\alpha + \beta)^{\frac{1}{3}} - (\alpha - \beta)^{\frac{1}{3}}]^2$$
$$= 4(x^{\frac{2}{3}} + y^{\frac{2}{3}}) = 4\,a^{\frac{2}{3}},$$
$$(\alpha + \beta)^{\frac{2}{3}} + (\alpha - \beta)^{\frac{2}{3}} = 2\,a^{\frac{2}{3}}.$$

6. If C is the centre of an ellipse, CG the X-intercept of the normal at P, and O the centre of curvature corresponding to P, prove that the area of the triangle COG is a maximum when the distance of P from the conjugate axis is one-fourth the transverse axis.

106. Envelopes. The equation of a locus is a relation between x, y, and one or more constants, upon which latter the position or form of the locus depends. Thus, the constants m and b fix the position of the straight line $y = mx + b$; the constants a and b determine the form of the ellipse $a^2y^2 + b^2x^2 = a^2b^2$; while the constants of the general equation of a conic determine both its position and form.

The constants in $y = f(x)$ are called **parameters**.

It follows that if different values be assigned to one of the parameters of the equation $y = f(x)$, the resulting equations will represent a series or system of curves differing from each other in form, or position, or both. Thus, $(x - m)^2 + y^2 = R^2$ is the equation of a circle whose centre is on X, and if different values be assigned to m, we shall obtain a series of equal circles whose centres are on X.

The curve which is tangent to all curves of the system obtained by the continuous variation of any one of the parameters in $y = f(x)$ is called the **envelope** *of the system.*

The constant thus supposed to vary is called the **variable parameter**.

Thus, in the case of the above circle $(x - m)^2 + y^2 = R^2$, m being the variable parameter, the envelope of the system is evidently the two tangents to the circle, in any of its positions, which lie parallel to X.

Denoting the variable parameter by a, the general equation of the system may be represented by

$$f(x, y, a) = 0.$$

107. Equation of the envelope.

Let SR be the envelope of any system of curves, and Q the point at which the envelope is tangent to any one curve of the system MN. Let

$$u = f(x, y, a) = 0 \qquad (1)$$

be the general equation of the system, a being the variable parameter, and P, (x, y), any point on MN.

Were MN fixed, that is, a constant, the direction of P's motion would be determined by

Fig. 32.

$$\frac{dy}{dx} = -\frac{\dfrac{du}{dx}}{\dfrac{du}{dy}}.$$

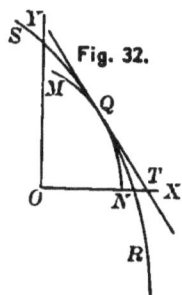

But, if MN is not fixed, a is variable, and

$$du = \frac{du}{dx} dx + \frac{du}{dy} dy + \frac{du}{da} da = 0,$$

whence

$$\frac{dy}{dx} = -\frac{\dfrac{du}{dx} + \dfrac{du}{da}\dfrac{da}{dx}}{\dfrac{du}{dy}}.$$

Now when P coincides with Q, these values of $\frac{dy}{dx}$ are equal, since MN and SR have at Q a common tangent. Hence at Q $\frac{du}{da}\frac{da}{dx} = 0$, which will be satisfied if $da = 0$, that is, if the partial derivative

$$\frac{du}{da} = 0. \qquad (2)$$

The coordinates of any point Q of the envelope must therefore satisfy (1) and (2). Hence, *to determine the equation of the envelope of any system, combine the general equation of the system with the equation formed by placing the partial derivative with respect to the variable parameter equal to zero, eliminating the parameter.*

EXAMPLES.   1. Find the envelope of $(x - m)^2 + y^2 = R^2$, m being the variable parameter.

$$u = (x - m)^2 + y^2 - R^2 = 0, \ \frac{du}{dm} = - 2(x - m) = 0, \text{ or } x = m.$$

Substituting this value of m in $(x - m)^2 + y^2 - R^2 = 0$ we have $y = \pm R$, two straight lines parallel to X.

2. Find the envelope of the hypothenuse of a right-angled triangle of constant area.

Let $OAB = c$ be the constant area, and $OA = a$. Then, since

$$\frac{OB \cdot OA}{2} = c, \quad OB = \frac{2c}{a}.$$

Hence the equation of AB is

$$\frac{x}{a} + \frac{y}{\dfrac{2c}{a}} = 1,$$

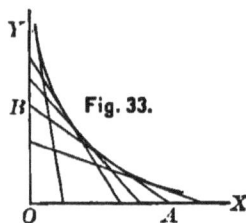

Fig. 33.

or
$$u = \frac{2cx}{a^2} + y - \frac{2c}{a} = 0, \tag{1}$$

in which a is the variable parameter. Hence

$$\frac{du}{da} = - \frac{4cx}{a^3} + \frac{2c}{a^2} = 0,$$

whence $a = 2x$. Substituting this value in (1), we have $xy = \dfrac{c}{2}$, the equation of an hyperbola referred to its asymptotes.

3. Find the envelope of an ellipse whose eccentricity so varies that its area remains constant; knowing the area of an ellipse to be πab.

We have $\pi ab = m$, a constant, whence $ab = \dfrac{m}{\pi} = c$, a constant. As a and b are both variable, we eliminate either parameter, as b, from $a^2y^2 + b^2x^2 = a^2b^2$ by means of the condition $ab = c$, and thus obtain $u = a^4y^2 + c^2x^2 - a^2c^2 = 0$; whence $\dfrac{du}{da} = 4\,a^3y^2 - 2\,ac^2 = 0$, or $a^2 = \dfrac{c^2}{2\,y^2}$, which in $a^4y^2 + c^2x^2 - a^2c^2 = 0$, gives $xy = \dfrac{c}{2}$. Since the axes are rectangular, the hyperbola is equilateral, as also in Ex. 2.

4. A line of fixed length moves with its extremities in two rectangular axes. Find its envelope.

Let AB (Fig. 33) be the line. Its equation is

$$\frac{x}{a} + \frac{y}{b} = 1, \quad \text{or} \quad u = bx + ay - ab = 0, \tag{1}$$

and by condition,

$$a^2 + b^2 = AB^2 = l^2, \tag{2}$$

l being constant. Proceeding as before, we should eliminate one of the parameters from (1) by means of (2) and then form the partial derivative. But it will be found more expeditious to differentiate first and eliminate afterwards.

We have from (1), since b is a function of a,

$$\frac{du}{da} = \frac{db}{da}(x - a) + y - b = -\frac{a}{b}(x - a) + y - b = 0, \tag{3}$$

since $\dfrac{db}{da} = -\dfrac{a}{b}$ from (2). Substituting in succession the values of x and y from (1) in (3), we find

$$a^2y + b^2y - b^3 = 0, \tag{4}$$

$$-a^2x - b^2x + a^3 = 0. \tag{5}$$

Substituting from (2) the value of a^2 in (4) and of b^2 in (5),

$$y = \frac{b^3}{l^2}, \quad x = \frac{a^3}{l^2},$$

or $\quad b^2 = y^{\frac{2}{3}}l^{\frac{4}{3}}, \quad a^2 = x^{\frac{2}{3}}l^{\frac{4}{3}},$

which in (2) give $x^{\frac{2}{3}} + y^{\frac{2}{3}} = l^{\frac{2}{3}}$, the equation of the hypocycloid.

5. Find the envelope of $y = mx + b$, m being the variable parameter.

6. From a point A on the axis of X distant a from the origin lines are drawn. Find the envelope of the perpendiculars drawn to these lines at their intersections with Y.

A line through A is $y = m(x - a)$, and its intersection with Y is $(0, -ma)$. The perpendicular to $y = m(x - a)$ at $(0, -ma)$ is $y + ma = -\dfrac{1}{m}x$. Hence $u = y + ma + \dfrac{x}{m} = 0$, in which m is the variable parameter.

$$\frac{du}{dm} = a - \frac{x}{m^2} = 0, \quad \therefore \ m = \frac{x^{\frac{1}{2}}}{a^{\frac{1}{2}}}.$$

Substituting this value of m in $y + ma + \dfrac{x}{m} = 0$, we have $y^2 = 4ax$, a parabola.

7. Find the envelope of a series of equal circles whose centres lie in the circumference of a given circle.

Let $x_1^2 + y_1^2 = R_1^2$ be the fixed circle. Then

$$(x - x_1)^2 + (y - y_1)^2 = R^2$$

is the movable circle.

Ans. $x^2 + y^2 = (R_1 \pm R)^2$, two concentric circles whose radii are $R_1 + R$ and $R_1 - R$.

8. Find the envelope of $x \cos 3\theta + y \sin 3\theta = a(\cos 2\theta)^{\frac{3}{2}}$, θ being the variable parameter.

$$x \cos 3\theta + y \sin 3\theta = a(\cos 2\theta)^{\frac{3}{2}}, \tag{1}$$

whence
$$\frac{du}{d\theta} = -x \sin 3\theta + y \cos 3\theta + a(\cos 2\theta)^{\frac{1}{2}} \sin 2\theta = 0,$$

or
$$x \sin 3\theta - y \cos 3\theta = a \sin 2\theta (\cos 2\theta)^{\frac{1}{2}}. \tag{2}$$

Squaring (1) and (2) and adding,

$$x^2 + y^2 = a^2[(\cos 2\theta)^3 + \cos 2\theta(\sin 2\theta)^2] = a^2 \cos 2\theta. \tag{3}$$

Dividing (2) by (1), ·

$$\frac{x \sin 3\theta - y \cos 3\theta}{x \cos 3\theta + y \sin 3\theta} = \frac{\tan 3\theta - \dfrac{y}{x}}{1 + \tan 3\theta \cdot \dfrac{y}{x}} = \tan 2\theta,$$

whence $\dfrac{y}{x} = \tan \theta$. Hence from (3),

$$x^2 + y^2 = a^2 \frac{1 - \tan^2 \theta}{1 + \tan^2 \theta} = a^2 \frac{x^2 - y^2}{x^2 + y^2},$$

or $(x^2 + y^2)^2 = a^2(x^2 - y^2)$, the lemniscate.

108. *The evolute is the envelope of the normals to the involute.*

Let (x', y') be any point P' of the involute, (a, β) the corresponding point Q of the evolute, and ϕ the angle made
by the normal or radius of curvature
$\rho = P'Q$ with X. Then for SQ and
SP' we have

$$a - x' = \rho \cos \phi, \quad \beta - y' = \rho \sin \phi,$$

or $\quad a = x' + \rho \cos \phi, \quad \beta = y' + \rho \sin \phi. \quad (1)$

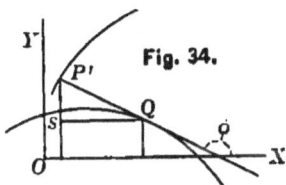

As (x', y') moves along the involute,
(a, β) moves along the evolute, or a, β, y' are functions of x'.
Hence, differentiating (1),

$$\left. \begin{array}{l} da = dx' + \cos \phi \, d\rho - \rho \sin \phi \, d\phi, \\ d\beta = dy' + \sin \phi \, d\rho + \rho \cos \phi \, d\phi. \end{array} \right\} \qquad (2)$$

But, Art. 26,

$$dx' = \sin \phi \, ds, \quad dy' = -\cos \phi \, ds,$$

or, since $\dfrac{1}{\rho} = \dfrac{d\phi}{ds}$,

$$dx' = \rho \sin \phi \, d\phi, \quad dy' = -\rho \cos \phi \, d\phi.$$

Substituting these in (2), we have

$$da = \cos \phi \, d\rho, \quad d\beta = \sin \phi \, d\rho, \qquad (3)$$

whence $\dfrac{d\beta}{da} = \tan \phi.$

But $\dfrac{d\beta}{da}$ is the slope of the tangent to the evolute at Q, and
$\tan \phi$ is the slope of the normal to the involute at P'.   Hence
*the normal to the involute is tangent to the evolute, and the evolute
is the envelope of the normals to the involute.*

109.   *The difference between any two radii of curvature to the
involute is equal to the arc of the evolute which they intercept.*

For, from Art. 108, Eq. 3,

$$da = \cos \phi d\rho, \quad d\beta = \sin \phi d\rho.$$

Hence, squaring and adding,

$$da^2 + d\beta^2 = d\rho^2;$$

or, if s' be the arc of the evolute (Art. 25),

$$ds' = \pm d\rho;$$

or the rates of change of s' and ρ are equal.

110.   The two preceding properties afford the following
mechanical construction of the involute when the evolute is
given.   Let RS be any curve.   Then,
if a pattern of RS be made, and a
string, one end of which is fixed at S,
be wrapped around the pattern SQR,
as the string is unwound from the
pattern the free end will describe the
curve MN which will be the involute of RS.   *Any* point of
the string will trace the arc of an involute as the string un-
winds from the evolute; hence, while a curve has but one
evolute, namely, the locus of its centre of curvature, the evolute has an infinite number of involutes.

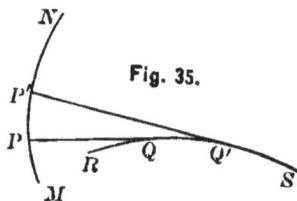

111. Orders of contact.

Let $y = f(x)$, $y = \phi(x)$, be the equations of two curves referred to the same axes and having a common point at $x = a$.

Then $f(a) = \phi(a)$. Let h be a very small increment of a, the ordinates corresponding to $x = a + h$ being $f(a+h)$, $\phi(a+h)$. By Taylor's formula,

$$f(a+h) = f(a) + f'(a)h + f''(a)\frac{h^2}{\underline{2}} + f'''(a)\frac{h^3}{\underline{3}} \cdots,$$

$$\phi(a+h) = \phi(a) + \phi'(a)h + \phi''(a)\frac{h^2}{\underline{2}} + \phi'''(a)\frac{h^3}{\underline{3}} \cdots,$$

or, by subtraction,

$$f(a+h) - \phi(a+h) = [f'(a) - \phi'(a)]h + [f''(a) - \phi''(a)]\frac{h^2}{\underline{2}}$$
$$+ [f'''(a) - \phi'''(a)]\frac{h^3}{\underline{3}} + \cdots, \quad (1)$$

which is the difference between corresponding ordinates of the curves on one or the other side of their common ordinate according as h is positive or negative. It thus appears from (1) that two curves are nearer on each side of their common point as the second member is smaller, that is, as the successive derivatives in order are equal each to each when $x = a$.

If $f'(a) = \phi'(a)$, the curves are tangent at $x = a$ and are said to have **contact of the first order**. If, also, $f''(a) = \phi''(a)$, the curves are said to have **contact of the second order**; and so on.

Cor. 1. Since, if the curves have a common point, we must have $f(a) = \phi(a)$, *contact of the nth order imposes $n + 1$ conditions.*

Cor. 2. If contact is of an odd order, the first term of (1) which does not vanish contains an even power of h, and the difference between the ordinates has the same sign whether h be positive or negative. Hence one curve lies above or below the other on both sides of the common ordinate, or *curves whose order of contact is odd do not intersect*. If contact is of an even order, the first term of (1) which does not vanish contains an odd power of h, and the difference between the ordinates changes sign with h. Hence if one curve lies above the other on one side of the common ordinate, it lies below it on the other side, or *curves whose order of contact is even intersect*.

Cor. 3. Since the number of independent conditions which can be imposed upon a curve is the same as the number of arbitrary constants in its equation, the highest possible order of contact between two curves whose general equations contain n and m arbitrary constants is $n - 1$, n being less than m.

EXAMPLES. 1. What is the highest possible order of contact of an ellipse and parabola?

The general equation of the conics contains five arbitrary constants, and therefore the ellipse has a possible fourth order of contact with other curves. But for the parabola $e = 1$, the number of arbitrary constants is four, and its highest possible order of contact is the third. Hence the ellipse and parabola cannot have contact with each other above the third order.

2. Prove that in general the highest possible order of contact of a straight line is the first, that is, tangency; and of the circle, the second.

3. Prove that at a point of inflexion the straight line has contact of the second order, and intersects the curve.

At a point of inflexion the second derivative of $y = f(x)$, the equation of the curve, is zero (Art. 96). Also, from $y = mx + b$, the second derivative is zero. Hence the line and the curve have contact of the second order. Hence, also, the tangent intersects the curve (Art. 111, Cor. 2).

4. Prove that at a point of maximum or minimum curvature the circle of curvature has contact of the third order.

At such a point the circle does not intersect the curve (Art. 103), hence its contact must be of an odd, and therefore of the third, order (Art. 111, Cor. 2).

SINGULAR POINTS.

112. Points of a curve presenting some peculiarity, independent of the position of the axes, are called **singular points**. Such are points of inflexion, already considered (Art. 103).

Multiple points. A multiple point is one common to two or more branches of a curve, and is double, triple, etc., as it lies on two, three, etc., branches.

If the branches pass *through* the point, as in Figs. 36 and 37, P is called a multiple point of **intersection** or **osculation**, according as the branches have different tangents or a common tangent. Thus, in Fig. 36, P is a triple multiple point of intersection; and in Fig. 37, P is a double multiple point of osculation.

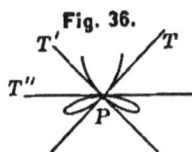

Fig. 36.

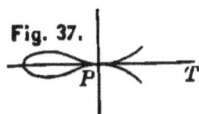

Fig. 37.

If the branches *meet* at the common point but do not pass through it, as in Figs. 38 and 39, P is called a **salient** point or a **cusp** point, according as the branches have different tangents or a common tangent. Cusp points are of the first or second species according as the branches lie on opposite sides or on the same side of the common tangent.

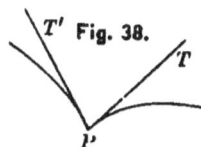

Fig. 38.

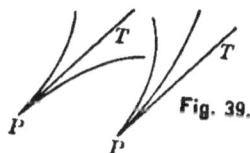

Fig. 39.

113. A **conjugate**, or **isolated point**, is one whose coordinates satisfy the equation of the curve, although no branch of the curve in the plane of the axes passes through it; as P, Fig. 40.

A **stop** point is one at which a single branch of a curve terminates.

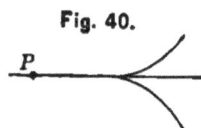

Fig. 40.

114. *Determination of singular points by inspection.*

Ascertain if possible, by inspection of the equation, whether for any value of one of the variables, as x, y has a single value. Let $x = a$ be the value of x which gives a single value b for y. Then the point (a, b) is to be examined.

If, for values of x both a little less and a little greater than a, y has more than one real value, the branches pass through

(a, b), which is therefore a multiple point of intersection or osculation. If, for values of x a little greater (or less) than a, y has more than one real value, but is imaginary when x is a little less (or greater) than a, the branches meet but do not pass through (a, b), which is therefore a salient or cusp point. If y is imaginary for values of x both a little less and a little greater than a, (a, b) is a conjugate point.

To determine whether the branches have the same tangent or different tangents at (a, b), we observe that, since (a, b) is common to several branches, $\frac{dy}{dx}$ must at that point have several values, and the branches will have different tangents or a common tangent according as these values are different or equal.

It is evident that $\frac{dy}{dx}$, as derived from $f(x, y) = 0$, may have more than one limit when $f(x, y) = 0$, has multiple points. Thus if POP' is the locus of $f(x, y) = 0$, $\frac{dy}{dx}$, being entirely general, applies to both the branch OP and the branch OP', and its value at O is the limit of $\frac{y}{x}$ or of $\frac{y'}{x'}$

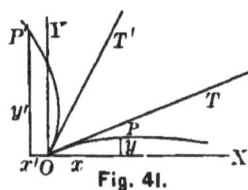

Fig. 41.

according as P or P' approaches O. It is thus a general expression for the limits of different ratios, and these limits may or may not be the same.

EXAMPLES. 1. Prove that $y^2 = x^2(1 - x^2)$ has a double multiple point of intersection at the origin.

Values of x, whether positive or negative, give in general two values of y; but when $x = 0$, y has the single value 0. Hence the branches pass through the orgin.

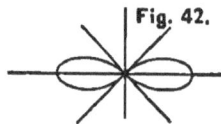

Fig. 42.

$$\frac{dy}{dx} = \frac{1 - 2x^2}{\sqrt{1 - x^2}}\bigg]_0 = \pm 1;$$

there are therefore two tangents at the origin, making angles of 45° and 135° with X, and the branches intersect.

2. Prove that $a^3y^2 - 2abx^2y - x^5 = 0$ has a point of osculation at the origin.

Solving f for y we have

$$y = \frac{x^2}{a^2}(b \pm \sqrt{ax + b^2}).$$

If x is positive and very small, the radical $> b$; hence one value of y is positive and the other nega-
tive. If x is negative and very small, both values of y are positive, since the radical is then less than b. If $x = 0$, $y = 0$. Hence the branches pass through the origin and lie in the second angle on the left of Y, and in the first and fourth on the right of Y.

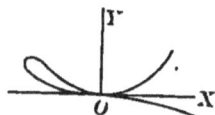

Fig. 43.

$$\frac{dy}{dx} = \frac{5x^4 + 4abxy}{2a^3y - 2abx^2}\Bigg]_{0,0} = \frac{20x^3 + 4aby + 4abx\dfrac{dy}{dx}}{2a^3\dfrac{dy}{dx} - 4abx}\Bigg]_{0,0} = \frac{0}{2a^3\dfrac{dy}{dx}},$$

whence $\frac{dy}{dx} = \pm 0$, or the axis of X is a common tangent at the origin. Hence there is a double point of osculation at the origin, and for one branch the origin is a point of inflexion.

Fig. 44.

3. Prove that $y^2 = 2x^2 + x^3$ has a multiple point of intersection at the origin, the tangent having the slopes $\pm\sqrt{2}$.

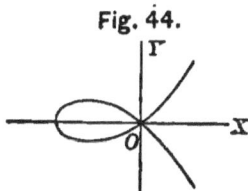

4. Prove that $y^2 = \frac{x^4}{a^2 - x^2}$ has a double multiple point of osculation at the origin.

Fig. 45.

y has in general two real values with opposite signs, whether x be positive or negative, and is zero when $x = 0$; hence the branches pass through the origin.

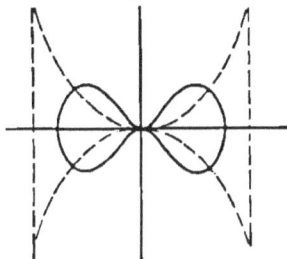

$$\frac{dy}{dx} = x\frac{2\,a^2 - x^2}{(a^2 - x^2)^{\frac{3}{2}}}\bigg]_0 = \pm\, 0.$$

Hence the axis of X is a common tangent at the origin.

5. Prove that $y^2 = x^4 - x^6$ has a double multiple point of osculation at the origin. The locus of Ex. 4 is represented by the dotted line of Fig. 45, and that of Ex. 5 by the full line.

6. Prove that the cissoid has a cusp of the first species at the origin.

$y^2 = \dfrac{x^3}{2\,a - x}.$ If x is positive, y has two values with opposite signs; if $x = 0$, $y = 0$; if x is negative, y is imaginary. Hence branches in the first and fourth angles meet at the origin, but do not pass through it, and the origin is either a salient point or a cusp of the first species.

$$\frac{dy}{dx} = x^{\frac{1}{2}}\frac{3\,a - x}{(2\,a - x)^{\frac{3}{2}}}\bigg]_0 = \pm\, 0,$$

or the branches have the axis of X for a common tangent.

7. Prove that $ay^2 = x^3$ has a cusp point of the first species at the origin.

8. Prove that $(y - x^2)^2 = x^5$ has a cusp point of the second species at the origin.

$y = x^2 \pm x^{\frac{5}{2}}.$ If x is negative, y is imaginary; if $x = 0$, $y = 0$; if x is positive and small, y has two positive values. Hence two branches, both in the first angle in the vicinity of the origin, meet at the origin but do not pass through it.

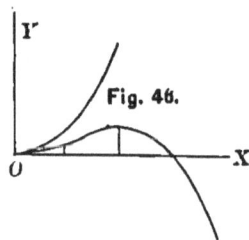

Fig. 46.

$$\frac{dy}{dx} = x(2 \pm \tfrac{5}{2}\sqrt{x})\big]_{x=0} = 0.$$

Hence X is a common tangent, and the origin is a cusp of the second species.

9. Show that $y^2 = x(a+x)^2$ has a conjugate point at $(-a, 0)$. $y = \sqrt{x}(a + x)$ has two values when x is positive, but is imaginary for all negative values of x except $x = -a$, when $y = 0$.

10. Prove that the conchoid has a multiple point of intersection, a cusp of the first species, or a conjugate point, at $(0, -b)$, according as $a > b$, $a = b$, $a < b$.

$$x^2y^2 = (y + b)^2(a^2 - y^2), \text{ whence } x = \pm \frac{y + b}{y}\sqrt{a^2 - y^2}.$$

If $a > b$, values of y a little less or greater than $-b$ give two values of x, and $x = 0$ when $y = -b$. Hence the branches pass through $(0, -b)$.

If $a = b$, x is imaginary if y is negative and numerically greater than b; is 0 when $y = -b$; and has two values when y is negative and numerically less than b. Hence the branches meet at $(0, -b)$, but do not pass through it.

If $a < b$, all negative values of y numerically greater than a, except $y = -b$, render x imaginary.

$$\frac{dy}{dx} = \frac{y^2x}{-x^2y + a^2y - 2y^3 + ba^2 - 3by^2 - b^2y}\Big]_{0,\,0}$$

$$= \frac{2yx\frac{dy}{dx} + y^2}{-2xy + (a^2 - x^2 - 6y^2 - 6by - b^2)\frac{dy}{dx}}\Big]_{0,\,0}$$

$$= \frac{b^2}{(a^2 - b^2)\frac{dy}{dx}}.$$

$$\therefore \frac{dy}{dx} = \pm \frac{b}{\sqrt{a^2 - b^2}}.$$

If $a > b$, there are two tangents who slopes are $\pm \dfrac{b}{\sqrt{a^2 - b^2}}$.

If $a = b$, the slope becomes ∞, and Y is a common tangent.

If $a < b$, $\dfrac{dy}{dx}$ is imaginary.

11. Show that $y = x \tan^{-1}\dfrac{1}{x}$ has a salient point at the origin.

If $x = 0$, $y = 0$; whether x be positive or negative, y is positive. The curve therefore lies above X, and branches in the first and second angles meet at the origin. When x is positive,

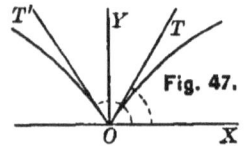

Fig. 47.

$$\frac{dy}{dx} = \tan^{-1}\frac{1}{x} - \frac{x}{1+x^2}\Big]_0 = \frac{\pi}{2} = 1.5708,$$

the slope of the branch in the first angle. When x is negative,

$$\frac{dy}{dx} = \tan^{-1}\left(-\frac{1}{x}\right) + \frac{x}{1+x^2}\Big]_0 = \tan(-\infty) = -\frac{\pi}{2} = -1.5708,$$

the slope of the branch in the second angle.

12. Prove that $y = x \log x$ has a stop point at the origin.

The curve lies to the right of Y, for negative numbers have no logarithms, and x cannot be negative. When x is positive, y has one real value. When $x = 0$,

Fig. 48.

$$y = x \log x = \frac{\log x}{\dfrac{1}{x}}\Big]_0 = -\frac{\dfrac{1}{x}}{\dfrac{1}{x^2}}\Big]_0 = 0.$$

Hence the curve consists of a single branch terminating at the origin.

ASYMPTOTES.

115. **A rectilinear asymptote** to a curve is a straight line which the curve continually approaches but never reaches; or it may be defined as the limiting position of the tangent as the point of contact recedes indefinitely from the origin.

If the curve has no infinite branch, it can have no asymptote.

116. Asymptotes parallel to the axes.

If PQ is an asymptote parallel to X, and at a distance b from it, then as x increases without limit, y approaches the finite limit b, and $y = b$ is the equation of PQ. So, also, if SR is an asymptote parallel to Y, and at a distance a from it, then as y increases without limit, x approaches the finite limit a, and $x = a$ is the equation of SR.

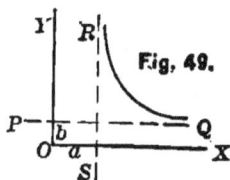

Fig. 49.

To determine, therefore, whether $f(x, y) = 0$ has asymptotes parallel to the axes, observe whether either variable approaches a finite value as a limit, that is, as the other increases indefinitely. If such be the case, there is an asymptote parallel to the axis corresponding to the variable which increases indefinitely, at a distance from it equal to the corresponding finite limit of the other variable.

EXAMPLES. 1. Show that $x = 2R$ is an asymptote to the cissoid.

$y^2 = \dfrac{x^3}{2R - x}$, in which y approaches $\pm \infty$ as x approaches $2R$. Hence $x = 2R$ is an asymptote to both branches.

2. Show that $y = 0$ is an asymptote to the conchoid.

$x = \pm \dfrac{y + b}{y} \sqrt{a^2 - y^2}$, in which, whether y be positive or negative, x approaches $\pm \infty$ as y approaches 0. Hence $y = 0$, or the axis of X, is an asymptote to both the branch above and that below X.

3. Examine $y = \tan x$ for asymptotes.

4. Show that $y = 0$ is an asymptote to the witch

$$x^2 y = 4R^2(2R - y).$$

5. $a^2y - x^2y = a^3$.

$y = \dfrac{a^3}{a^2 - x^2}$. As x approaches $\pm \infty$, y approaches 0. Hence $y = 0$, or the axis of X, is an asymptote to two branches. Also, y approaches ∞ as x approaches $\pm a$. Hence $x = a$ and $x = -a$ are asymptotes.

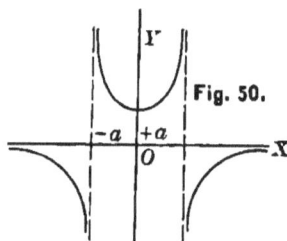

Fig. 50.

6. $a^2x = y(x - a)^2$.

$y = \dfrac{a^2x}{(x - a)^2}$. As x approaches $\pm \infty$, y approaches ± 0. Also y approaches ∞ as x approaches a. Hence the axis of X and $x = a$ are asymptotes.

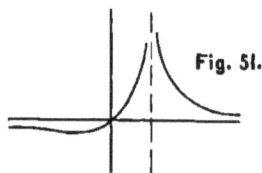

Fig. 51.

7. $xy - ay - bx = 0$.

$y = \dfrac{bx}{x - a}$, $x = \dfrac{ay}{y - b}$. The asymptotes are $x = a$, $y = b$.

8. Show that $y = 0$ is an asymptote to $x = \log y$.

9. Examine $x^2y^2 = a^2(x^2 - y^2)$ for asymptotes. *Ans.* $y = \pm a$.

10. Examine $y(a^2 + x^2) = a^2(a - x)$ for asymptotes.

Ans. $y = 0$.

11. Examine $y = a + \dfrac{b^3}{(x - c)^2}$ for asymptotes.

Ans. $y = a$, $x = c$.

12. Examine the locus of Ex. 4, Art. 114, for asymptotes.

117. Asymptotes oblique to the axes.

The equation of a tangent to a plane curve being

$$y - y' = \frac{dy'}{dx'}(x - x'),$$

if we make in this equation $y = 0$, the corresponding value of

x will be the intercept of the tangent on X. Representing this intercept by X, we have

$$X = x' - y' \frac{dx'}{dy'}. \tag{1}$$

In like manner, making $x = 0$, the intercept on Y is found to be

$$Y = y' - x' \frac{dy'}{dx'}, \tag{2}$$

from which the accents may be omitted if we understand (x, y) is the point of tangency. Now the asymptote is the limiting position of the tangent, that is, the position which the tangent approaches as the point of contact recedes indefinitely; hence its slope is the limit of $\frac{dy}{dx}$, and its intercepts are the limits of (1) and (2), as the point of contact recedes indefinitely from the origin. The position of the asymptote when oblique to the axes will therefore be known when the limits of X and Y are known, and if these limits be designated by X_1 and Y_1, the equation of the asymptote is

$$\frac{x}{X_1} + \frac{y}{Y_1} = 1.$$

If either X_1 or Y_1 is zero, the asymptote passes through the origin, and its direction is determined by finding the limit of $\frac{dy}{dx}$ as the point of contact recedes indefinitely from the origin. If both X_1 and Y_1 are infinity, there is no asymptote. If one is infinity and the other finite or zero, the asymptote is parallel to or coincides with the axis on which the intercept is infinite.

It is usually most expeditious to find first the limit of $\frac{dy}{dx}$. If this is neither 0 nor ∞, the asymptote is oblique, and its position is made known by either X_1 or Y_1; if the limit of $\frac{dy}{dx}$ is zero, there will be an asymptote parallel to the axis of X if Y_1 is finite; if the limit of $\frac{dy}{dx}$ is ∞, there will be an asymptote parallel to the axis of Y if X_1 is finite.

EXAMPLES.   Examine the following curves for asymptotes.

1. The parabola, $y^2 = 2px$.

The curve has infinite branches in the first and fourth angles.

$\dfrac{dy}{dx} = \dfrac{p}{y}\bigg]_{\pm\infty} = \pm\, 0$; hence if there are asymptotes, they are parallel to the axis of X. $Y_1 = y - x\dfrac{dy}{dx}\bigg]_{\infty,\,\pm\infty} = \dfrac{y}{2}\bigg]_{\pm\infty} = \pm\, \infty$; hence there are no asymptotes.

2. The hyperbola, $a^2y^2 - b^2x^2 = -a^2b^2$.

The curve has an infinite branch in each angle.

$$\frac{dy}{dx} = \frac{b^2x}{a^2y} = \pm\,\frac{b}{a}\sqrt{1 + \frac{b^2}{y^2}}\bigg]_{\pm\infty} = \pm\,\frac{b}{a}.$$

$$X_1 = x - y\frac{dx}{dy}\bigg]_{\infty,\,\infty} = \frac{a^2}{x}\bigg]_{\pm\infty} = \pm\, 0.$$

Hence the diagonals of the rectangle on the axes are asymptotes to the curve in each angle.

3. $x = \log y$.

Since y approaches 0 as x approaches $-\infty$, the axis of X is an asymptote (Art. 142).   Otherwise, $\dfrac{dy}{dx} = y\bigg]_0 = 0$; hence if there is an asymptote to the branch in the second angle, it is parallel to X.

$$Y_1 = y - y\log y]_0 = \frac{1 - \log y}{\dfrac{1}{y}}\Bigg]_0 = \frac{\dfrac{1}{y}}{-\dfrac{1}{y^2}}\Bigg]_0 = 0,$$

or the axis of X is the asymptote.

For the branch in the first angle, $x = \infty$ when $y = \infty$. Hence $\dfrac{dy}{dx} = y\bigg]_\infty = \infty$; that is, the asymptote is perpendicular to the axis of X, if one exists. $X_1 = x - 1]_\infty = \infty$, or there is no asymptote to the branch in the first angle.

4. $y^3 = x^2(a - x)$.

When $x > a$, y is negative, and there is an infinite branch in the fourth angle. When x is negative, y is positive, and there is also an infinite branch in the second angle.

$$X_1 = \frac{-ax}{2a - 3x}\bigg]_{x=\infty} = \frac{a}{3}. \qquad Y_1 = \frac{a}{3\left(\dfrac{a}{x} - 1\right)^{\frac{1}{3}}}\bigg]_{x=\infty} = \frac{a}{3}.$$

Hence the asymptote is common to both branches, and its equation is $y = -x + \dfrac{a}{3}$. (See Fig. 53.)

5. Prove that $y = x + 2$ is an asymptote to $y^3 = 6x^2 + x^3$.

6. Prove that $y = -x$ is an asymptote to $y^3 = a^3 - x^3$.

CURVE TRACING.

118. The foregoing principles are sufficient for the determination of the forms and singularities of many curves, but a knowledge of the general theory of curves is necessary in order to trace curves with facility from their equations.

1. $y = \dfrac{x}{1 + x^2}$ (1), $\qquad f'(x) = \dfrac{1 - x^2}{(1 + x^2)^2}$ (2),

$$f''(x) = \frac{2x(x^2 - 3)}{(1 + x^2)^3} \quad (3).$$

Since y has but one value for any value of x, its sign being that of x, and is 0 when $y = 0$, the curve passes from the third to the first angle through the origin, and has infinite branches in these angles. As x approaches $\pm \infty$, y approaches 0, and the axis of X is therefore an asymptote to both branches. $f'(x)$ changes sign at

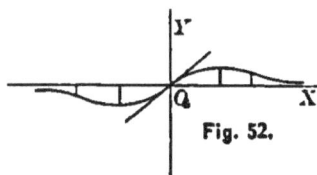

Fig. 52.

$x = \pm 1$, and these values render $f''(x)$ negative and positive respectively, giving a maximum ordinate in the first, and a minimum ordinate in the third, angle. $f''(x)$ changes sign at

$x = \pm \sqrt{3}$ and $x = 0$, giving three points of inflexion. The slope of the curve at the origin is $45°$, for $f'(x) = 1$ when $x = 0$.

2. $y^3 = ax^2 - x^3$ (1), $\qquad f'(x) = \dfrac{2a - 3x}{3x^{\frac{1}{3}}(a - x)^{\frac{2}{3}}}$ (2),

$$f''(x) = -\frac{2a^2}{9x^{\frac{4}{3}}(a-x)^{\frac{5}{3}}} \quad (3).$$

If x is negative, y is positive, and there is an infinite branch in the second angle. $f''(x)$ is negative when x is negative, hence this branch is convex upward.

If x is positive, y is positive till $x = a$, when $y = 0$, the curve having a branch which crosses X at $x = a$ from the first into the fourth angle. Since $y = 0$ when $x = 0$, the branches meet at the origin, which is a cusp point of the first species, $f'(x)$ becoming ∞ for $x = 0$, and Y being the common tan-

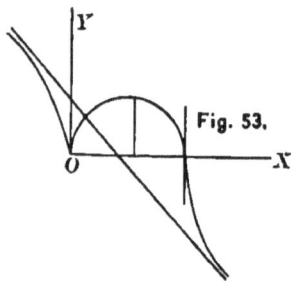

Fig. 53.

gent. $f''(x)$ changes sign from $-$ to $+$ at $x = a$, which is therefore a point of inflexion, the curve being convex upward in the first angle and concave in the fourth. The slope at $x = a$ is ∞, since $f'(x) = \infty$ when $x = a$. $f'(x)$ changes sign at $x = \frac{2}{3}a$ from $+$ to $-$, hence $x = \frac{2}{3}a$ renders y a maximum. It has been shown in Art. 145, Ex. 4, that $y = -x + \dfrac{a}{3}$ is an asymptote to both branches.

3. $y = e^{-\frac{1}{x}}$ (1), $\quad f'(x) = \dfrac{1}{x^2 e^{\frac{1}{x}}}$ (2), $\quad f''(x) = -\dfrac{2x-1}{x^4 e^{\frac{1}{x}}}$ (3).

From (1) we observe that y is positive whatever the value of x, or the curve lies above X.

Let x be negative. Then $y = e^{\frac{1}{x}}$, which increases as x decreases, becoming ∞ when $x = 0$; and decreases as x increases,

becoming 1 when $x = \infty$. Hence $y = 1$, and
the axis of Y, are asymptotes in the second
angle. Also, when x is negative, $f''(x)$ is
positive, and this branch is concave upward.

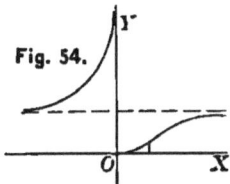

Fig. 54.

Let x be positive. Then $y = \dfrac{1}{e^{\frac{1}{x}}}$, which

increases with x, becoming 1 when $x = \infty$; or, $y = 1$ is also an
asymptote to the branch in the first angle. Since $y = 0$ when
$x = 0$, and $y = \infty$ when x is negative and very small, the origin
is a stop point.

$f'(x)$ cannot change sign, hence there are neither maxima nor
minima ordinates.

$f''(x)$ changes at $x = \frac{1}{2}$ from $+$ to $-$, a point of inflexion at
which the curvature changes from concave to convex upward.

$f'(x) = \dfrac{1}{x^2 e^{\frac{1}{x}}} \Big]_0 = 0 \times \infty$. Placing $z = \dfrac{1}{x}$, whence $z = \infty$ when

$x = 0$, $f'(x) = \dfrac{z^2}{e^z}\Big]_\infty = \dfrac{2z}{e^z}\Big]_\infty = \dfrac{2}{e^z}\Big]_\infty = 0$, and X is a tangent at the
origin.

4. $y = x \log x$ (1), $f'(x) = 1 + \log x$ (2), $f''(x) = \frac{1}{x}$ (3).

The curve lies to the right of Y, since x cannot be negative.
As the logarithm of a proper fraction is negative, y is negative
till $x = 1$, when $y = 0$. When $x > 1$, y is
positive. As $f''(x)$ cannot change sign,
the curve is concave upward. $f'(x) = 0$
gives $\log x = -1$, or $x = e^{-1} = \dfrac{1}{e}$, which
renders y a minimum. When $x = 0$,
$f'(x) = -\infty$, or the axis of Y is a tangent. When $x = 1$, $f'(x) = 1$, or the curve crosses X at an
angle of $45°$.

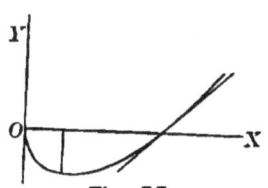

Fig. 55.

$$X_1 = x - y\frac{dx}{dy} = \frac{x}{1 + \log x}\Big]_\infty = \frac{1}{\dfrac{1}{x}}\Big]_\infty = \infty,$$

$$Y_1 = y - x\frac{dy}{dx} = -x\Big]_\infty = \infty,$$

hence there are no asymptotes. The origin is a stop point (Art. 114, Ex. 11).

5. $(y - x)^2 = x^5$, or $y = x^2 \pm x^{\frac{5}{2}}$ (1), $f'(x) = x(2 \pm \frac{5}{2}\sqrt{x})$ (2),
$$f''(x) = 2 \pm \tfrac{15}{4}\sqrt{x} \ (3).$$

See Ex. 7, Art. 114, and Fig. 46.

6. $y^3 = a^2x$ (1). $\quad f'(x) = \dfrac{a^2}{3\,y^2}$ (2), $\quad f''(x) = -\dfrac{2\,a^4}{9\,y^5}$ (3).

7. $ay^2 = x^3$ (1), $\quad f'(x) = \dfrac{3\,x^2}{2\,ay}$ (2), $\quad f''(x) = \dfrac{3\,x^4}{4\,a^2y^3}$ (3).

8. $y^2 = 2\,x^2 + x^3$ (1), $\quad f'(x) = \dfrac{4\,x + 3\,x^2}{2\,y} = \pm\dfrac{4 + 3\,x}{2\sqrt{2 + x}}$ (2),
$$f''(x) = \pm \dfrac{8 + 3\,x}{4(2 + x)^{\frac{3}{2}}} \ (3).$$

From (1), $y = \pm x\sqrt{2 + x}$, from which we see that the curve is symmetrical to X, passes through the origin, and has $x = -2$, $x = \infty$ for its limits along X. $f'(x) = \pm\sqrt{2}$ when $x = 0$, hence the origin is a multiple point of intersection. The tangent at $x = -2$ is perpendicular to X, since $f'(x) = \infty$ for $x = -2$. $f''(x)$ has two signs, but does not change sign except for $x = -\frac{8}{3}$, which is not a point of the curve, since the limit of x is -2; hence there are no points of inflexion.

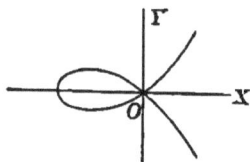

Fig. 56.

$f'(x)$ changes sign at $x = -\frac{4}{3}$, where there is a maximum and a minimum ordinate.

$$f'(x) = \dfrac{dy}{dx} = \dfrac{4\,x + 3\,x^2}{2\,y}\Bigg]_{\substack{x = \infty \\ y = \infty}} = \dfrac{4 + 6\,x}{2\dfrac{dy}{dx}}, \quad \text{whence } \dfrac{dy}{dx}\Bigg]_{x = \infty} = \infty,$$

or the asymptote, if there be one, is perpendicular to X. $X_1 = \dfrac{x^2}{4 + 3\,x}\Bigg]_{\infty} = \infty$, and there is no asymptote.

9. $y^2 = x^4 + x^5$ (1), $f'(x) = \dfrac{x^3}{2y}(4 + 5x) = \pm\dfrac{4x + 5x^2}{2\sqrt{1 + x}}$ (2),

$$f''(x) = \pm\frac{8 + 24x + 15x^2}{4(1 + x)^{\frac{3}{2}}} \ (3).$$

From (1), $y = \pm x^2\sqrt{1 + x}$, whence we see the curve is symmetrical to X, passes through the origin, and that its limits along X are $x = -1$ and $x = \infty$. When $x = 0$, $f'(x) = \pm 0$; hence the origin is a point of osculation, X being a tangent to both branches.

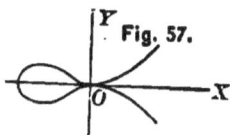

Fig. 57.

From $f''(x) = 0$, we find $x = \dfrac{-12 \pm \sqrt{24}}{15}$; the lower sign is impossible since $x = -1$ is a limit, and the upper sign gives points of inflexion. $f'(x) = 0$, gives $x = 0$ and $x = -\frac{4}{5}$, which correspond to maxima and minima ordinates. There are no asymptotes.

10. $y^2 = x^3 - x^4$, or $y = \pm x^{\frac{3}{2}}\sqrt{1 - x}$ (1),

$$f'(x) = \frac{3x^2 - 4x^3}{2y} = \pm\frac{x}{2}\frac{3 - 4x}{\sqrt{x(1 - x)}} \ (2),$$

$$f''(x) = \pm\frac{8x^2 - 12x + 3}{4(1 - x)\sqrt{x(1 - x)}} \ (3).$$

From (1) the curve is seen to be symmetrical with respect to X; and as x cannot be negative and $f'(x) = 0$ when $x = 0$, the origin is a cusp of the first species. Since x cannot be greater than 1, the curve lies between the limits 0 and 1 along X. There is a maximum and a minimum ordinate at $x = \frac{3}{4}$,

Fig. 58.

and $x = \dfrac{3 - \sqrt{3}}{4}$ corresponds to points of inflexion. When $x = 1$, $f'(x) = \infty$.

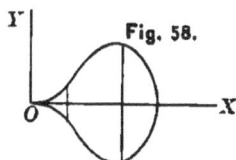

11. $a^2y - x^2y = a^3$.

12. $4x = y(x - 2)^2$.

13. $ay^2 - x^3 + bx^2 = 0$.

14. $x^3 - xy + 1 = 0$.

15. $y^2 = x^2(1 - x^2)^3$ (Fig. 59).   18. $y^2 = x^4 - x^6$ (Fig. 45).

16. $y^2 = x^4(1 - x^2)^3$ (Fig. 60).   19. $y^3 + x^2 - a^3 = 0$.

17. $a^3y^3 + b^3x^3 = a^3b^3$.

Fig. 59.

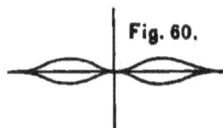
Fig. 60.

POLAR CURVES.

119. Subtangent and subnormal.

Let P be any point of MM', PT the tangent, PN the normal, O the pole, and OX the polar axis. Through the pole draw NT perpendicular to the radius vector to the point of contact, OP, meeting the tangent and the normal at T and N. Then OT is the subtangent and ON the subnormal.

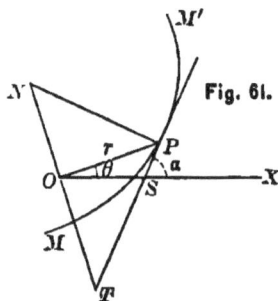
Fig. 61.

120. Lengths of the subtangent and tangent.

From the right triangle OPT,

$$OT = OP \tan OPT = r \tan OPT.$$

But $\tan OPT = \tan (\alpha - \theta) = \dfrac{\tan \alpha - \tan \theta}{1 + \tan \alpha \tan \theta}$

$$= \frac{\dfrac{dy}{dx} - \tan \theta}{1 + \dfrac{dy}{dx} \tan \theta} = \frac{dy \cos \theta - dx \sin \theta}{dx \cos \theta + dy \sin \theta}.$$

But $x = r \cos \theta$, $y = r \sin \theta$, whence

$$dx = \cos \theta \, dr - r \sin \theta \, d\theta, \quad dy = \sin \theta \, dr + r \cos \theta \, d\theta.$$

Making these substitutions, we find $\tan OPT = r \dfrac{d\theta}{dr}$; hence

$$\text{Subtangent} = OT = r^2 \frac{d\theta}{dr}.$$

and $\qquad \text{Tangent} = \sqrt{OP^2 + OT^2} = r\sqrt{1 + r^2\dfrac{d\theta^2}{dr^2}}.$

Cor. $\qquad ds = \sqrt{dx^2 + dy^2}$ (Art. 25) $= \sqrt{dr^2 + r^2 d\theta^2}.$

121. Lengths of the subnormal and normal.

$$ON = OP\tan OPN = r\cot OPT = \frac{r}{\tan OPT}.$$

Hence (Art. 120),

$$\text{Subnormal} = ON = \frac{dr}{d\theta},$$

and $\qquad \text{Normal} = PN = \sqrt{r^2 + \dfrac{dr^2}{d\theta^2}}.$

EXAMPLES. 1. The lemniscate $r^2 = a^2\cos 2\theta$.

$$\frac{dr}{d\theta} = -\frac{a^2\sin 2\theta}{r}.$$

Hence $\qquad \text{Subt} = r^2\dfrac{d\theta}{dr} = -\dfrac{r^3}{a^2\sin 2\theta} = -r\cot 2\theta,$

$$\text{Tangent} = r\sqrt{1 + r^2\frac{d\theta^2}{dr^2}} = \frac{ra^2}{\sqrt{a^4 - r^4}},$$

$$\text{Subn} = \frac{dr}{d\theta} = -\frac{a^2\sin 2\theta}{r},$$

$$\text{Normal} = \sqrt{r^2 + \frac{dr^2}{d\theta^2}} = \frac{a^2}{r}.$$

2. The logarithmic spiral $r = a^\theta$.

$$\frac{dr}{d\theta} = a^\theta \log a.$$

Hence $\qquad \text{Subt} = \dfrac{r}{\log a}.$ $\qquad \text{Subn} = r\log a.$

In this spiral $\tan OPT = r\dfrac{d\theta}{dr} = \dfrac{1}{\log a}$, a constant. Hence the tangent makes a constant angle with the radius vector to the point of contact. For this reason this spiral is often called the equiangular spiral. In the Naperian logarithmic spiral, $\log e = 1$, and the subnormal is equal to the radius vector.

3. The spiral of Archimedes $r = a\theta$.

$$\frac{dr}{d\theta} = a = \text{subn, a constant.} \quad r^2\frac{d\theta}{dr} = a\theta^2 = \text{subt.}$$

4. The cardioide $r = a(1 + \cos\theta)$.

$$\frac{dr}{d\theta} = -a\sin\theta = \text{subn.} \quad \text{Subt.} = -\frac{r^2}{a\sin\theta}.$$

5. Prove that in the curve $r = a\sin\theta$ the radius vector makes equal angles with the tangent and polar axis.

$$\tan OPT = r\frac{d\theta}{dr} = a\sin\theta\frac{1}{a\cos\theta} = \tan\theta.$$

6. The circle $r = 2R\cos\theta$.

$$\text{Subt} = 2R\cot\theta\cos\theta, \quad \text{subn} = -2R\sin\theta,$$
$$\text{Tangent} = 2R\cos\theta\,\text{cosec}\,\theta, \quad \text{normal} = 2R.$$

7. Prove that the subtangent of the reciprocal spiral is constant.

122. Curvature of polar curves.

A curve at any of its points is said to be convex or concave towards the pole according as its tangent does or does not lie on the same side of the curve as the pole.

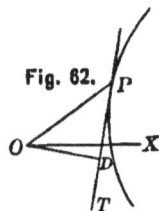

Fig. 62.

Let fall from the pole the perpendicular $OD = p$ upon the tangent. If the curve is concave to the pole, p is an increasing function of r, $r = f(\theta)$ being the equation of the curve. Therefore $\frac{dp}{dr}$ is positive. If the curve is convex to the pole, p is a decreasing function of r, and $\frac{dp}{dr}$ is negative.

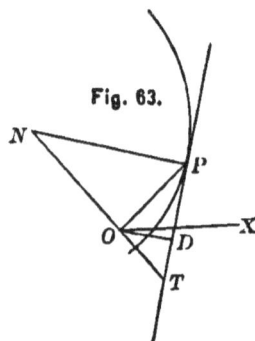

Fig. 63.

Hence the curve is concave or convex towards the pole according as $\dfrac{dp}{dr}$ is positive or negative, and at a point of inflexion $\dfrac{dp}{dr}$ must change sign.

To find p, we have (Fig. 63), NP being the normal,

$$OD : NP :: OT : NT.$$

But, Arts. 120 and 121,

$$NP = \sqrt{r^2 + \frac{dr^2}{d\theta^2}}, \qquad OT = r^2 \frac{d\theta}{dr},$$

$$NT = NO + OT = \frac{dr}{d\theta} + r^2 \frac{d\theta}{dr}.$$

Hence $\qquad p = \dfrac{r^2}{\sqrt{r^2 + \dfrac{dr^2}{d\theta^2}}}.$

To examine a polar curve for points of inflexion, substitute $\dfrac{dr}{d\theta}$ from the equation of the curve, $r = f(\theta)$, in the above value of p, and see if, for any value of r, $\dfrac{dp}{dr}$ changes sign.

EXAMPLES. 1. Prove that the logarithmic spiral is always concave to the pole.

$$r = a^\theta, \quad \therefore \ \frac{dr}{d\theta} = r \log a, \quad p = \frac{r^2}{\sqrt{r^2 + \dfrac{dr^2}{d\theta^2}}} = \frac{r}{\sqrt{1 + \log^2 a}}.$$

Hence $\qquad \dfrac{dp}{dr} = \dfrac{1}{\sqrt{1 + \log^2 a}},$

which is always positive.

2. Examine the lituus for curvature.

$$r = \frac{a}{\sqrt{\theta}}, \quad \therefore \ \frac{dr}{d\theta} = -\frac{a}{2\theta^{\frac{3}{2}}}, \quad p = \frac{2a^2 r}{\sqrt{4a^4 + r^4}}.$$

Hence $\qquad \dfrac{dp}{dr} = \dfrac{2a^2(4a^4 - r^4)}{(4a^4 + r^4)^{\frac{3}{2}}} = 0,$

gives $r = a\sqrt{2}$. If $r < a\sqrt{2}$, $\dfrac{dp}{dr}$ is positive; if $r > a\sqrt{2}$, $\dfrac{dp}{dr}$ is negative. Hence $r = a\sqrt{2}$ indicates a point of inflexion at which the curvature changes from concave to convex towards the pole.

3. Prove that the parabola $r = \dfrac{p'}{1 - \cos\theta}$ is always concave to the pole.

123. Radius of curvature.

From Art. 119, we have

$$\rho = \frac{\left[1 + \left(\dfrac{dy}{dx}\right)^2\right]^{\frac{3}{2}}}{\dfrac{d^2y}{dx^2}}, \qquad (1)$$

in which x is equicrescent, and the problem is to transform (1) into its equivalent in terms of r and θ when θ is equicrescent.

Therefore (Art. 58, Ex. 7),

$$\rho = \frac{\left[\left(\dfrac{dr}{d\theta}\right)^2 + r^2\right]^{\frac{3}{2}}}{r^2 + 2\left(\dfrac{dr}{d\theta}\right)^2 - r\dfrac{d^2r}{d\theta^2}}. \qquad (2)$$

EXAMPLES.　Find the radius of curvature of:

1. The lemniscate, $r^2 = a^2 \cos 2\theta$.

$$\frac{dr}{d\theta} = -\frac{a^2 \sin 2\theta}{r} = -\frac{\sqrt{a^4 - r^4}}{r},$$

$$\frac{d^2r}{d\theta^2} = -\frac{r^4 + a^4}{r^3}.$$

Hence　　　　　　$\rho = \dfrac{a^2}{3r}.$

2. The cardioide, $r = a(1 - \cos \theta)$.

$$\frac{dr}{d\theta} = a \sin \theta = (2ar - r^2)^{\frac{1}{2}}.$$

$$\frac{d^2r}{d\theta^2} = a \cos \theta = a - r.$$

Hence $\qquad \rho = \frac{2}{3}\sqrt{2ar}.$

3. The spiral of Archimedes, $r = a\theta$.

$$\rho = \frac{(r^2 + a^2)^{\frac{3}{2}}}{2a^2 + r^2} = a\frac{(1 + \theta^2)^{\frac{3}{2}}}{2 + \theta^2}.$$

4. The reciprocal spiral, $r = \dfrac{a}{\theta}$.

$$\rho = \frac{r(a^2 + r^2)^{\frac{3}{2}}}{a^3} = \frac{a}{\theta^4}(1 + \theta^2)^{\frac{3}{2}}.$$

5. The lituus, $r = \dfrac{a}{\sqrt{\theta}}$.

$$\rho = \frac{r}{2a^2}\frac{(4a^4 + r^4)^{\frac{3}{2}}}{4a^4 - r^4}.$$

6. The logarithmic spiral, $r = a^\theta$.

$$\rho = a^\theta(1 + \log^2 a)^{\frac{1}{2}}.$$

If $a = e$, $\rho = \sqrt{2}e^\theta = r\sqrt{2}$, or the radius of curvature is $\sqrt{2} \times$ the radius vector.

124. Asymptotes.

Since the asymptote is the limiting position of the tangent as the point of contact recedes indefinitely from the pole, if a polar curve has an asymptote, r must be infinite for some finite value of θ, and for such value of θ the subtangent $r^2\dfrac{d\theta}{dr}$ must be finite.

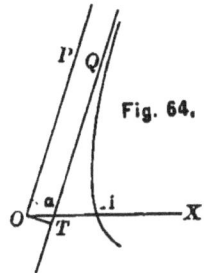

Fig. 64.

Let a be the value of θ which renders r infinity. To construct the asymptote make $AOP = a$, draw through O a perpendicular to

OP, and make $OT = r^2 \dfrac{d\theta}{dr}\bigg]_{r=a}$. Then TQ, parallel to OP, is the asymptote. For the point of contact being infinitely distant from O, the radius vector and asymptote are parallel. If $r^2 \dfrac{d\theta}{dr}\bigg]_{\theta=a} = \infty$, there is no asymptote.

EXAMPLES. 1. Examine the hyperbola for asymptotes.

$$r = \frac{p}{e\cos\theta - 1}, \text{ whence } \frac{dr}{d\theta} = \frac{ep\sin\theta}{(e\cos\theta - 1)^2} \text{ and } r^2\frac{d\theta}{dr} = \frac{p}{e\sin\theta}.$$

Now $r = \infty$ when $\cos\theta = \dfrac{1}{e} = \dfrac{a}{\sqrt{a^2 + b^2}}$. Hence, if there be an asymptote, it is parallel to the diagonal of the rectangle on the axes. Again,

$$\sin\theta = \sqrt{1 - \cos^2\theta} = \frac{\sqrt{e^2 - 1}}{e},$$

hence $r^2\dfrac{d\theta}{dr} = \dfrac{p}{e\sin\theta} = a\sqrt{e^2 - 1} = b.$

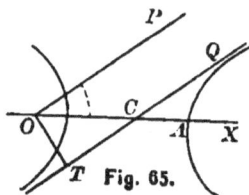
T Fig. 65.

There is therefore an asymptote. To construct it, draw OP parallel to the diagonal on the axes $\left(\text{or make } AOP = \cos^{-1}\dfrac{1}{e}\right)$, and make $OT = b$. Then TQ, parallel to OP, is the asymptote. Since $OC = \dfrac{OT}{\sin a} = \dfrac{be}{\sqrt{e^2 - 1}} = ae$, C is the centre, and the asymptote coincides with the diagonal. Also, as

$$\cos\theta = \cos(-\theta),$$

there is another asymptote below the axis, and similarly situated.

2. Prove that the parabola $r = \dfrac{p}{1 - \cos\theta}$ has no asymptote.

3. Prove that the lituus $r = \dfrac{a}{\sqrt{\theta}}$ has the polar axis for an asymptote.

4. Prove that the spiral $r = \dfrac{a}{\theta}$ has an asymptote parallel to the polar axis at a distance a from it.

5. Examine $r^2 = \dfrac{a^2 \sin 3\theta}{\cos \theta}$ for asymptotes.

6. Examine $(r - a) \sin \theta = b$ for asymptotes.

7. Examine the conchoid $r = b \operatorname{cosec} \theta + a$ for asymptotes.

$r = \infty$ when $\theta = 0$. $\quad r^2 \dfrac{d\theta}{dr} = \left. \dfrac{(b + a \sin \theta)^2}{b \cos \theta} \right]_0 = b.$

Hence the asymptote is parallel to the polar axis, and at a distance from it equal to b.

125. Tracing of polar curves.

Write the equation of the curve $f(r, \theta) = 0$ in the form $r = f(\theta)$, when possible, and assign such values to θ as will render it easy to determine those of r. This will usually be sufficient to determine the general form of the locus. For maxima or minima values of r, $\dfrac{dr}{d\theta}$ must change sign. The locus may then be examined for curvature, points of inflexion, and asymptotes.

EXAMPLES. 1. $r = a \sin 2\theta.$ $\quad \dfrac{dr}{d\theta} = 2a \cos 2\theta.$

$r = 0$ when $\theta = 0$ and increases with θ till $\theta = \dfrac{\pi}{4}$, when $\dfrac{dr}{d\theta}$ changes sign from $+$ to $-$. Hence $\theta = \dfrac{\pi}{4}$ renders $r = a$ a maximum.

From $\theta = \dfrac{\pi}{4}$ to $\theta = \dfrac{\pi}{2}$ r decreases from a to 0, and the curve is a loop in the first angle.

When θ passes $\dfrac{\pi}{2}$, r becomes negative, increasing numerically till $\theta = \frac{3}{4}\pi$, when $\dfrac{dr}{d\theta}$ changes sign from $-$ to $+$. Hence $\theta = \frac{3}{4}\pi$

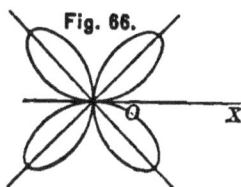

Fig. 66.

renders $r = -a$ a minimum. From $\theta = \frac{3}{4}\pi$ to $\theta = \pi$, r is still negative, but decreases numerically from a to 0, giving another loop in the fourth angle.

When θ passes π, r becomes positive, and as $\sin 2\theta$ passes through all its values while θ varies from 0 to π, equal loops will be traced for values of θ between π and 2π.

The maximum and minimum values of r are derived from $\frac{dr}{d\theta} = 0$; namely, $\frac{1}{4}\pi$, $\frac{3}{4}\pi$, $\frac{5}{4}\pi$, and $\frac{7}{4}\pi$.

No value of θ renders $r = \infty$; hence the curve has no asymptote.

2. $r = a \sin 3\theta$.

The curve is shown in the figure. From this example and Ex. 1 it may be inferred that in all equations of the form

$$r = a \sin n\theta,$$

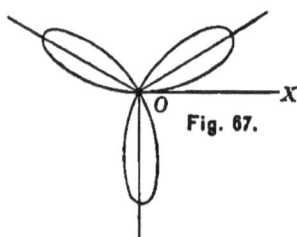

Fig. 67.

the curve consists of n, or of $2n$, loops, according as n is an odd or an even integer.

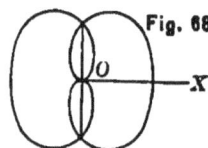

Fig. 68.

3. $r = a \sin \dfrac{\theta}{2}$ (Fig. 68).

4. $r = a(1 - \tan \theta)$ (Fig. 69).

5. $r = a \cos 2\theta$.

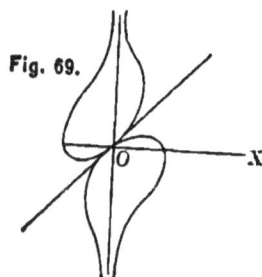

Fig. 69.

6. $r = a + \sin \frac{1}{2}\theta$.

7. $r = a + \sin \frac{2}{3}\theta$ (Fig. 70).

8. $r = a + \tan 2\theta$.

9. $r^2 = a^2(\tan^2 \theta - 1)$.

Prove that there are two asymptotes perpendicular to the polar axis at distances $\pm a$ from the pole.

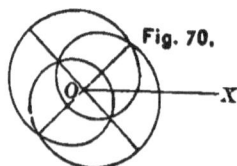

Fig. 70.

PART II.

THE INTEGRAL CALCULUS.

CHAPTER VI.

TYPE INTEGRABLE FORMS.

126. **Integral and Integration.**

If $f(x)$ be any function, and $f'(x)dx$ its differential, then $f(x)$ is called the **integral** of $f'(x)dx$. Hence, *any function is the integral of its differential.*

The process of finding the function from its differential is called **integration**. As an operation it is the inverse of differentiation, and having seen

I. *Differentiation to be the process of finding the ratio of the rates of change of the function and its variable,* we may define

II. *Integration to be the process of finding the function when the ratio of its rate of change to that of its variable is given.*

127. **Symbol of integration.** The symbol of integration is $\int$, read 'the integral of.' Thus, if

$$y = x^3,$$

then
$$dy = 3x^2dx.$$

and
$$\int dy = y = \int 3x^2dx = x^3,$$

d and $\int$, as symbols of inverse operations, neutralizing each other.

The test of the result of any integration is differentiation; that is, $\int 3x^2dx = x^3$ because $d(x^3) = 3x^2dx.$

128. Constant of integration.

It is evident that functions which have the same rate, and therefore the same differential, may differ from each other by any constant, but only by a constant. Thus in the function $y = mx + b$, which for different values of b represents a series of parallel straight lines, the rate of y will be the same whatever the value of b, or $dy = mdx$ for all values of b. Hence any given differential is the differential of an infinite number of functions which differ from each other by a constant, and if the differential only is known, the function cannot be determined. Therefore

$$\int mdx = mx + C,$$

in which C is an undetermined constant.

Otherwise: since the differential of a constant is zero, if a function contains any constant *term*, this term will not appear in its differential; hence a constant C must be added to every integral to represent this term.

This constant is called the constant of integration. It will be shown, in the application of integration to definite problems, that it may either be eliminated or that its value may be determined from the conditions.

129. *The integral of the sum of any number of terms is the sum of the integrals of the terms.*

This is an obvious consequence of the proposition (Art. 16) that the differential of the sum of any number of terms is the sum of their differentials.

Or, formally, as $\int$ and d neutralize each other,

$$\int d(x - y + z) = x - y + z + C,$$

and $$\int dx - \int dy + \int dz = x - y + z + C;$$

hence
$$\int d(x - y + z) = \int dx + \int dy + \int dz.$$

Both d and $\int$ are, therefore, *distributive* symbols.

- **130.** *If the differential has a constant factor, its integral will have the same constant factor.*

For
$$d[Af(x)] = Ad[f(x)], \tag{1}$$

and
$$\int d[Af(x)] = Af(x); \tag{2}$$

but (2) is the integral of (1).

Since a constant factor in the differential also appears in the integral, *such a factor may be written before or after the integral sign, at pleasure.*

Thus, $d(ax) = adx$, and

$$\int adx = a\int dx = ax.$$

131. Type integrable forms.

Since a function is the integral of its differential, from $d(ax^m) = max^{m-1}dx$, we have

$$\int max^{m-1}dx = ax^m,$$

or
$$\int ax^{m-1}dx = \frac{ax^m}{m}.$$

Putting $m - 1 = n$, we have in general,

$$\int ax^n dx = \frac{a}{n+1}x^{n+1} + C.$$

Reversing the fundamental processes of differentiation, we obtain thus the twenty following forms:

1. $\displaystyle\int ax^n dx = \frac{a}{n+1}x^{n+1} + C.$

2. $\displaystyle\int \frac{dx}{x} = \log x + C.$

3. $\int a^x \log a \, dx = a^x + C.$

4. $\int e^x dx = e^x + C.$

5. $\int \cos x \, dx = \sin x + C.$

6. $\int - \sin x \, dx = \cos x + C.$

7. $\int \sec^2 x \, dx = \tan x + C.$

8. $\int - \csc^2 x \, dx = \cot x + C.$

9. $\int \sec x \tan x \, dx = \sec x + C.$

10. $\int - \csc x \cot x \, dx = \csc x + C.$

11. $\int \sin x \, dx = \text{vers } x + C.$

12. $\int - \cos x \, dx = \text{covers } x + C.$

13. $\int \dfrac{dx}{\sqrt{1 - x^2}} = \sin^{-1} x + C.$

14. $\int - \dfrac{dx}{\sqrt{1 - x^2}} = \cos^{-1} x + C.$

15. $\int \dfrac{dx}{1 + x^2} = \tan^{-1} x + C.$

16. $\int - \dfrac{dx}{1 + x^2} = \cot^{-1} x + C.$

17. $\int \dfrac{dx}{x\sqrt{x^2 - 1}} = \sec^{-1} x + C.$

18. $\int - \dfrac{dx}{x\sqrt{x^2 - 1}} = \csc^{-1} x + C.$

19. $\int \dfrac{dx}{\sqrt{2x - x^2}} = \text{vers}^{-1} x + C.$

20. $\int - \dfrac{dx}{\sqrt{2x - x^2}} = \text{covers}^{-1} x + C.$

132. Remarks on the type forms.

The processes of the Integral Calculus consist chiefly in the reduction of differentials to the above forms. When this reduction has been effected, the integral is seen at once by inspection. This being the case, it is evidently indispensable that the student should be thoroughly familiar with the type forms, so as to be able to recognize them at sight. The following suggestions will facilitate their recognition and application.

FORM 1. *Whenever a differential can be resolved into three factors, viz.: a constant factor, a variable factor with any constant exponent except* − 1, *and a differential factor which is the differential of the variable factor without its exponent, then its integral is the product of the constant factor into the variable factor with its exponent increased by 1, divided by the new exponent.*

For $\qquad \int a \cdot x^n \cdot dx = \dfrac{a}{n+1} x^{n+1} + C.$

FORM 2. When the exponent of the variable factor is − 1, the differential falls under the second form

$$\int \frac{dx}{x} = \log x + C,$$

in which the numerator is the differential of the denominator. Hence, *whenever the numerator of a fraction is the differential of its denominator, the integral of the fraction is the Naperian logarithm of its denominator.*

FORMS 3 AND 4. These forms are

$$\int a^x \cdot \log a\, dx = a^x + C,$$

3

and $$\int e^x \cdot dx = e^x + C,$$

in which *the differential factor must be the logarithm of the base into the differential of the exponent.*

Forms 5–12. In each of these forms *the differential factor dx must be the differential of the arc.*

Forms 13–20. The conditions to which the differential must conform should in each case be carefully noted. Thus, from

$$\int \frac{dx}{1 + x^2} = \tan^{-1}x + C,$$

we see that the first term of the denominator must be 1, and the numerator the differential of the square root of the second term of the denominator.

Examples. 1. $\int 5x^2 dx = \int 5 \cdot x^2 \cdot dx = \frac{5}{3}x^3 + C.$ (Form 1.)

2. $\int mx^{-m}dx = \frac{m}{1 - m} x^{1-m} + C.$

3. $\int \frac{adx}{x^3} = \int ax^{-3}dx = -\frac{a}{2x^2} + C.$

4. $\int \frac{2}{3} \frac{dx}{x^{\frac{1}{3}}} = x^{\frac{2}{3}} + C.$

5. $\int \frac{dx}{\sqrt{x}} = 2\sqrt{x} + C.$

6. $\int \left(ax^2 - \frac{x^5}{b} + \sqrt{x} \right) dx = \frac{a}{3}x^3 - \frac{x^6}{6b} + \frac{2}{3}x^{\frac{3}{2}} + C.$ (Art. 129.)

7. $\int b(a+bx)^4 dx = \int (a+bx)^4 \cdot bdx = \frac{1}{5}(a+bx)^5 + C.$ (Form 1.)

8. $\int \frac{3x^2 dx}{(4 + x^3)^{\frac{1}{2}}} = \int (4 + x^3)^{-\frac{1}{2}} 3x^2 dx = 2(4 + x^3)^{\frac{1}{2}} + C.$

9. $\int m(3ax^2 + 5x^5)^{\frac{2}{3}}(6ax + 25x^4)dx = \frac{3}{5}m(3ax^2 + 5x^5)^{\frac{5}{3}} + C.$

10. $\int \dfrac{2\,x\,dx}{x^2+1} = \log(x^2+1)+C.$ (Form 2.)

In logarithmic integrals it is customary to write the constant of integration $C = \log c$.   Hence

$$\log(x^2+1)+C = \log(x^2+1)+\log c = \log[c(x^2+1)].$$

11. $\int \dfrac{dx}{x \pm a} = \log[c(x \pm a)].$

12. $\int \dfrac{3\,ax^2dx}{ax^3+b} = \log[c(ax^3+b)].$

13. $\int \dfrac{3}{2}\dfrac{1+x^2}{2+3x+x^3}\,dx = \dfrac{1}{2}\int \dfrac{3+3x^2}{2+3x+x^3}\,dx$
$$= \log[c(2+3x+x^3)^{\frac{1}{2}}].$$

14. $\int \dfrac{1+\cos x}{x+\sin x}\,dx = \log[c(x+\sin x)].$

15. $\int \dfrac{dx}{x \log x} = \int \dfrac{\frac{dx}{x}}{\log x} = \log(\log x) + \log c = \log[c \log x].$

16. $\int 10 \log^3 x\,\dfrac{dx}{x} = \frac{5}{2}\log^4 x + C.$ (Form 1.)

17. $\int m \log^n x\,\dfrac{dx}{x} = \dfrac{m}{n+1}\log^{n+1} x + C.$

18. $\int ae^{ax}dx = \int e^{ax}\cdot a\,dx = e^{ax} + C.$ (Form 4.)

19. $\int 3 \log a\, a^{x^3} x^2 dx = \int a^{x^3}\cdot \log a\, 3\, x^2 dx = a^{x^3} + C.$ (Form 3.)

20. $\int e^{\sin x}\cos x\,dx = e^{\sin x} + C.$

21. $\int \sin x \cos x\,dx = \frac{1}{2}\sin^2 x + C.$ (Form 1.)

22. $\int -2\sin 2x\,dx = \int -\sin 2x \cdot 2\,dx = \cos 2x + C.$

23. $\int 4\sin^3 x \cos x dx = \sin^4 x + C.$

24. $\int 4\sec^4 x \tan x dx = \int 4 \cdot \sec^3 x \cdot \sec x \tan x dx = \sec^4 x + C.$

25. $\int \frac{1}{2}\tan^4 x \sec^2 x dx = \frac{1}{10}\tan^5 x + C.$

26. $\int 6\tan x^3 \sec^2 x^3 \cdot x^2 dx = \int 2 \cdot \tan x^3 \sec^2 x^3 \cdot 3 x^2 dx$
$$= \tan^2 x^3 + C.$$

27. $\int \frac{2 x dx}{\sqrt{1 - x^4}} = \sin^{-1} x^2 + C.$

28. $\int \frac{2\sin^{-1} x dx}{\sqrt{1 - x^2}} = \int 2 \cdot \sin^{-1} x \cdot \frac{dx}{\sqrt{1 - x^2}} = (\sin^{-1} x)^2 + C.$
$$\text{(Form 1.)}$$

29. $\int \frac{5 dx}{\sqrt{10 x - 25 x^2}} = \int \frac{5 dx}{\sqrt{2(5 x) - (5 x)^2}} = \mathrm{vers}^{-1} 5 x + C.$

30. $\int \frac{dx}{x^2 + 4 x + 5} = \int \frac{dx}{1 + (x - 2)^2} = \tan^{-1}(x - 2) + C.$

31. $\int e^{e^x} e^x dx = e^{e^x} + C.$

ELEMENTARY TRANSFORMATIONS.

No general method exists for the reduction of differentials to type forms. Much therefore depends upon the ingenuity and insight of the student. In addition to the specific transformations applicable to certain differentials of definite forms, given in the next chapter, the following elementary transformations should constantly be borne in mind.

133. By the introduction of a constant factor. When the differential is under a type form *so far as the variable is concerned*, it may frequently be reduced exactly to such form by multiplying and dividing by a constant factor. This reduction depends upon the fact that a constant factor may be written before or after the integral sign.

ILLUSTRATIONS. FORM 1. $\int (3ax^2 + 2x)^3 \cdot (3ax + 1)dx.$

Were the differential factor $(6ax + 2)dx$, it would be the exact differential of the variable factor without its exponent. Hence, multiplying and dividing by 2,

$$\int (3ax^2 + 2x)^3(3ax + 1)dx = \tfrac{1}{2}\int (3ax^2 + 2x)^3(6ax + 2)dx$$
$$= \tfrac{1}{8}(3ax^2 + 2x)^4 + C.$$

When the proper factor is not readily seen by inspection, we may determine it as follows. Suppose the differential to be $(2x^{\frac{3}{2}} + x^5)^{\frac{3}{4}}(\tfrac{6}{5}x^{\frac{1}{2}} + 2x^4)dx$, and $A = $ required constant factor. Then A must satisfy the condition

$$d(2x^{\frac{3}{2}} + x^5) = (\tfrac{6}{5}Ax^{\frac{1}{2}} + 2Ax^4)dx,$$
or $\qquad (3x^{\frac{1}{2}} + 5x^4)dx = (\tfrac{6}{5}Ax^{\frac{1}{2}} + 2Ax^4)dx;$

and as this condition must be fulfilled for all values of x, the coefficients of like powers in the two members must be equal, or $\tfrac{6}{5}A = 3$, $2A = 5$, from either of which we find $A = \tfrac{5}{2}$. Introducing this factor,

$$\tfrac{2}{5}\int (2x^{\frac{3}{2}} + x^5)^{\frac{3}{4}}(3x^{\frac{1}{2}} + 5x^4)dx = \tfrac{8}{35}(2x^{\frac{3}{2}} + x^5)^{\frac{7}{4}} + C.$$

Again, suppose the given differential to be

$$(2x^3 + 7x)^{\frac{1}{2}}(5x^2 + 3)dx.$$

Then we must have

$$d(2x^3 + 7x) = (6x^2 + 7)dx = (5Ax^2 + 3A)dx;$$

whence $6 = 5A$, and $7 = 3A$, or $A = \tfrac{6}{5}$, $A = \tfrac{7}{3}$. As these values are not the same, there is no constant factor, and the integration cannot be effected by Form 1.

FORM 2. $\int \dfrac{3 + 2\,x^3}{6\,x + x^4}\,dx.$

Were the numerator $6 + 4\,x^3$, it would be the exact differential of the denominator. Hence, introducing the factor 2,

$$\int \frac{3 + 2\,x^3}{6\,x + x^4}\,dx = \tfrac{1}{2} \int \frac{6 + 4\,x^3}{6\,x + x^4} = \tfrac{1}{2}\log(6\,x + x^4) + \log c$$
$$= \log\left[c(6\,x + x^4)^{\frac{1}{2}}\right].$$

If the constant factor is not readily seen, it may be determined from the condition that the numerator must be the exact differential of the denominator.

FORMS 3 AND 4. The constant is determined from the condition that the differential factor must be the product of the logarithm of the base into the differential of the exponent. Thus, to integrate $a^{2x}dx$, the factor to be introduced is $2 \log a$, and

$$\int a^{2x}\,dx = \frac{1}{2 \log a} \int a^{2x} \cdot \log a\, 2\,dx = \frac{1}{2 \log a}\, a^{2x} + C.$$

FORMS 5 TO 12. The required constant is readily seen from the fact that the differential factor is the differential of the arc. Thus

$$\int \cos 2x\,dx = \tfrac{1}{2} \int \cos 2\,x \cdot 2\,dx = \tfrac{1}{2} \sin 2\,x + C.$$

FORMS 13 TO 20. In the case of the circular differentials the constant must be determined separately for each form. For example, having given $\int \dfrac{dx}{\sqrt{a^2 - b^2 x^2}}$, we observe that so far as the variable is concerned it has the type form $\int \dfrac{dx}{\sqrt{1 - x^2}}$. To transform it, we must make the first term under the radical 1, and the numerator the exact differential of the square root of the second term under the radical. We proceed, therefore, as follows :

$$(13') \quad \int \frac{dx}{\sqrt{a^2 - b^2 x^2}} = \int \frac{\frac{1}{a}\, dx}{\sqrt{1 - \frac{b^2}{a^2} x^2}} = \frac{1}{b} \int \frac{\frac{b}{a}\, dx}{\sqrt{1 - \frac{b^2}{a^2} x^2}}$$
$$= \frac{1}{b} \sin^{-1} \frac{b}{a} x + C.$$

$$(14') \quad \int - \frac{dx}{\sqrt{a^2 - b^2 x^2}} = \frac{1}{b} \cos^{-1} \frac{b}{a} x + C.$$

$$(15') \quad \int \frac{dx}{a^2 + b^2 x^2} = \frac{1}{a} \int \frac{\frac{1}{a}\, dx}{1 + \frac{b^2}{a^2} x^2} = \frac{1}{ab} \int \frac{\frac{b}{a}\, dx}{1 + \frac{b^2}{a^2} x^2}$$
$$= \frac{1}{ab} \tan^{-1} \frac{b}{a} x + C.$$

$$(16') \quad \int - \frac{dx}{a^2 + b^2 x^2} = \frac{1}{ab} \cot^{-1} \frac{b}{a} x + C.$$

$$(17') \quad \int \frac{dx}{x\sqrt{b^2 x^2 - a^2}} = \frac{1}{a} \int \frac{dx}{x\sqrt{\frac{b^2}{a^2} x^2 - 1}} = \frac{1}{a} \int \frac{\frac{b}{a}\, dx}{\frac{b}{a} x \sqrt{\frac{b^2}{a^2} x^2 - 1}}$$
$$= \frac{1}{a} \sec^{-1} \frac{b}{a} x + C.$$

$$(18') \quad \int - \frac{dx}{x\sqrt{b^2 x^2 - a^2}} = \frac{1}{a} \operatorname{cosec}^{-1} \frac{b}{a} x + C.$$

$$(19') \quad \int \frac{dx}{\sqrt{2abx - b^2 x^2}} = \int \frac{\frac{1}{a}\, dx}{\sqrt{\frac{2b}{a} x - \frac{b^2}{a^2} x^2}} = \frac{1}{b} \int \frac{\frac{b}{a}\, dx}{\sqrt{\frac{2b}{a} x - \frac{b^2}{a^2} x^2}}$$
$$= \frac{1}{b} \operatorname{vers}^{-1} \frac{b}{a} x + C.$$

$$(20') \quad \int - \frac{dx}{\sqrt{2abx - b^2 x^2}} = \frac{1}{b} \operatorname{covers}^{-1} \frac{b}{a} x + C.$$

These eight forms are known as the subordinate, or auxiliary, circular forms. It is better to transform each special case directly; thus

$$\int \frac{4\,dx}{3+5x^2} = \frac{4}{3}\int \frac{dx}{1+\frac{5}{3}x^2} = \frac{4}{3}\frac{\sqrt{3}}{\sqrt{5}}\int \frac{\sqrt{\frac{5}{3}}\,dx}{1+\frac{5}{3}x^2}$$

$$= \frac{4}{\sqrt{15}}\tan^{-1}\sqrt{\tfrac{5}{3}}\,x + C.$$

If the subordinate forms are memorized, or at hand for reference, we have

$$\frac{4\,dx}{3+5x^2} = \frac{dx}{\frac{3}{4}+\frac{5}{4}x^2},$$

whence, by comparison with (15′),

$$a^2 = \tfrac{3}{4},\ \ b^2 = \tfrac{5}{4}.$$

and hence

$$\frac{1}{ab}\tan^{-1}\frac{b}{a}x + C = \frac{4}{\sqrt{15}}\tan^{-1}\sqrt{\tfrac{5}{3}}\,x + C.$$

EXAMPLES.

1. $\displaystyle \int \tfrac{2}{3}(x^5+1)^{\frac{3}{2}}x^4dx = \tfrac{2}{3}\cdot\tfrac{1}{5}\int (x^5+1)^{\frac{3}{2}}5x^4dx = \tfrac{4}{15}(x^5+1)^{\frac{1}{2}}+C.$

2. $\displaystyle \int \left(1+\frac{9a}{4}x\right)^{\frac{1}{2}}dx = \frac{8}{27a}(1+\tfrac{9}{4}ax)^{\frac{3}{2}} + C.$

3. $\displaystyle \int \sqrt{\frac{2r}{2r-y}}\,dy = -2\sqrt{2r}\sqrt{2r-y} + C.$

4. $\displaystyle \int \sqrt{2x^4 - 3x^2 + 1}\,(x^3 - \tfrac{3}{4}x)dx = \tfrac{1}{12}(2x^4 - 3x^2+1)^{\frac{3}{2}}+C.$

5. Which of the following can be integrated by introducing a constant factor?

$$(5x^4 + 3x^3 + x^2 + 5)^{\frac{2}{3}}(2x^3 + \tfrac{9}{10}x^2 + \tfrac{1}{5}x)dx.$$

$$(3x^2 - 2x)^{\frac{2}{3}}(3x - 1)dx.$$

$$\frac{3x^2 - 1}{(1 - x + x^3)^2}\,dx. \qquad \frac{4 + 6x^2}{(4x - 3x^3)^2}\,dx.$$

6. $\int \dfrac{ax^2 - \frac{1}{2}bx^{\frac{1}{2}}}{ax^3 - bx^{\frac{3}{2}}}\,dx = \dfrac{1}{3}\int \dfrac{3ax^2 - \frac{3}{2}bx^{\frac{1}{2}}}{ax^3 - bx^{\frac{3}{2}}}\,dx$

$\qquad = \log\left[c\left(ax^3 - bx^{\frac{3}{2}}\right)^{\frac{1}{3}}\right].$

7. $\int \dfrac{1}{4}\dfrac{9\,x^2 dx}{12\,x^3 + 7} = \log\left[c(12\,x^3 + 7)^{\frac{1}{16}}\right].$

8. $\int \dfrac{x^2 dx}{a - bx^3} = \log\dfrac{c}{(a - bx^3)^{\frac{1}{3b}}}.$

9. $\int \dfrac{5\,x^2 dx}{10\,x^3 + 16} = \log\left[c(10\,x^3 + 16)^{\frac{1}{6}}\right].$

10. $\int \dfrac{\sin x\,dx}{a + b\cos x} = \log\dfrac{c}{(a + b\cos x)^{\frac{1}{b}}}.$

11. Which of the following can be integrated by the introduction of a constant factor?

$$\dfrac{5\,dx}{8 - 6x^2}. \qquad \dfrac{1 - \sqrt{x}}{x - 7\,x^{\frac{3}{2}}}\,dx. \qquad \dfrac{x^{m-1}dx}{x^m + 1} \qquad \dfrac{4x - 3\sqrt{x}}{x^2 - x^{\frac{3}{2}}}\,dx.$$

12. $\int a^{\alpha x}dx = \dfrac{1}{\alpha \log a}\,a^{\alpha x} + C.$

13. $\int a^{x^3}x^2 dx = \dfrac{1}{3\log a}\,a^{x^3} + C.$

14. $\int me^{4x}dx = \frac{3}{2}me^{4x} + C.$

15. $\int e^{\cos x}\sin x\,dx = -\,e^{\cos x} + C.$

16. $\int e^{\sin \frac{x}{2}}\cos\dfrac{x}{2}\,dx = 2\,e^{\sin \frac{x}{2}} + C.$

17. $\int \dfrac{e^{2\tan^{-1}x}}{1 + x^2}\,dx = \frac{1}{2}e^{2\tan^{-1}x} + C.$

18. Which of the following can be integrated by introducing a constant factor?

$$e^x\,a\log a\,\dfrac{dx}{x^2}. \qquad e^{x^n}\,n\,dx. \qquad e^x dx.$$

19. $\int \cos^3 x \sin x dx = -\frac{1}{4}\cos^4 x + C.$

20. $\int \sin^2 4x \cos 4x dx = \frac{1}{12}\sin^3 4x + C.$

21. $\int \sin x^3 \cdot x^2 dx = -\frac{1}{3}\cos x^3 + C.$

22. $\int \frac{2}{3}\sec^2 x^3 \tan x^3 \cdot x^2 dx = \frac{1}{4}\sec^2 x^3 + C.$

23. $\int \frac{dx}{\sqrt{4 - 9x^2}} = \frac{1}{3}\sin^{-1}\frac{3}{2}x + C.$

24. $\int -\frac{dx}{\sqrt{a(b^2 - x^2)}} = \frac{1}{\sqrt{a}}\cos^{-1}\frac{x}{b} + C.$

25. $\int \frac{4\,dx}{1 + 4x^2} = 2\tan^{-1}2x + C.$

26. $\int \frac{x^3 dx}{1 + x^8} = \frac{1}{4}\tan^{-1}x^4 + C.$

27. $\int \frac{3\,dx}{5 + 7x^2} = \frac{3}{\sqrt{35}}\tan^{-1}\sqrt{\frac{7}{5}}x + C.$

28. $\int \frac{2\,dx}{x\sqrt{3x^2 - 5}} = \frac{2}{\sqrt{5}}\sec^{-1}\sqrt{\frac{3}{5}}x + C.$

29. $\int \frac{dx}{\sqrt{2ax - x^2}} = \text{vers}^{-1}\frac{x}{a} + C.$

134. By the transference of a variable factor. Although a variable factor cannot be taken out from under the integral sign, it may be transferred from one factor of the differential expression to another, or introduced into both terms of a fraction.

ILLUSTRATION.

$$\int \frac{3}{2}(ax^{13} + x^{15})^{\frac{1}{2}}(5a + 7x^2)dx = \frac{3}{2}\int (ax^5 + x^7)^{\frac{1}{2}}(5ax^4 + 7x^6)dx$$
$$= (ax^5 + x^7)^{\frac{3}{2}} + C.$$

EXAMPLES.

1. $\displaystyle\int\frac{2\,a^2+4\,x^2}{\sqrt{a^2+x^2}}dx=\int\frac{2\,a^2x+4\,x^3}{\sqrt{a^2x^2+x^4}}dx=2\,(a^2x^2+x^4)^{\frac12}+C.$

2. $\displaystyle\int\frac{dx}{(1-x^2)^{\frac32}}=\int\frac{x^{-3}dx}{(x^{-2}-1)^{\frac32}}=-\frac{x}{\sqrt{1-x^2}}+C.$

3. $\displaystyle\int\frac{xdx}{(8\,x+x^2)^{\frac32}}=\int\frac{\dfrac{dx}{x^2}}{\left(\dfrac{8}{x}+1\right)^{\frac32}}=-\int\left(\frac{8}{x}+1\right)^{-\frac32}\left(-\frac{dx}{x^2}\right)$

$\qquad\qquad\qquad=\dfrac14\dfrac{x}{\sqrt{8\,x+x^2}}+C.$

4. $\displaystyle\int\frac{2\,x^2+1}{x^3+x}dx=\log\left[c\,(x^4+x^2)^{\frac12}\right].$

5. $\displaystyle\int\frac{dx}{x\sqrt{x^2-a^2}}=\int\frac{dx}{\sqrt{x^4-a^2x^2}}=-\frac{1}{a}\int\frac{-a\dfrac{dx}{x^2}}{\sqrt{1-\dfrac{a^2}{x^2}}}$

$\qquad\qquad\qquad=-\dfrac1a\sin^{-1}\dfrac{a}{x}+C.$

6. $\displaystyle\int\frac{dx}{e^{-x}+e^x}=\int\frac{e^x dx}{1+e^{2x}}=\tan^{-1}e^x+C.$

7. $\displaystyle\int\frac{3}{4}\frac{dx}{x^{\frac14}(1+x^{\frac32})}=\int\frac{\frac34 x^{-\frac14}dx}{1+x^{\frac32}}=\tan^{-1}x^{\frac34}+C.$

135. By expansion. When the exponents of the factors of the differential are positive integers, the indicated operations may be performed, and the resulting monomial terms integrated separately. Care should be taken not to expand unnecessarily; thus,

$$\int(1-x)^2dx=-\tfrac13(1-x)^3+C.$$

EXAMPLES.

1. $\displaystyle\int(1+x^2)(1-x)dx=x-\frac{x^2}{2}+\frac{x^3}{3}-\frac{x^4}{4}+C.$

2. $\int (x+1)^3 x^2 dx = \dfrac{x^6}{6} + \tfrac{3}{5} x^5 + \tfrac{3}{4} x^4 + \tfrac{1}{3} x^3 + C.$

3. $\int (a+bx)^2 x dx = \dfrac{a^2}{2} x^2 + \dfrac{2ab}{3} x^3 + \dfrac{b^2}{4} x^4 + C.$

4. $\int (1 - x + x^2)^2 dx = x - x^2 + x^3 - \tfrac{1}{2} x^4 + \tfrac{1}{5} x^5 + C.$

5. $\int \dfrac{a^2}{2} (1 + \sin 4x) dx = \dfrac{a^2}{2} (x - \tfrac{1}{4} \cos 4x) + C.$

6. $\int (a - x^2)^3 x^{\frac{1}{2}} dx = \dfrac{2a^3}{3} x^{\frac{3}{2}} - \dfrac{6a^2}{7} x^{\frac{7}{2}} + \dfrac{6a}{11} x^{\frac{11}{2}} - \tfrac{2}{15} x^{\frac{15}{2}} + C.$

136. By division. Expansion by division will often lead to integration, as may be seen by the following

EXAMPLES.

1. $\int \dfrac{x^4 dx}{1 + x^2} = \int \left(x^2 - 1 + \dfrac{1}{1 + x^2} \right) dx = \tfrac{1}{3} x^3 - x + \tan^{-1} x + C.$

2. $\int \dfrac{x^5 + x}{1 + x^2} dx = \tfrac{1}{4} x^4 - \tfrac{1}{2} x^2 + \log(1 + x^2) + C.$

3. $\int \dfrac{x^2 + 1}{x - 1} dx = \tfrac{1}{2} x^2 + x + \log(x - 1)^2 + C.$

4. $\int \dfrac{(3 + \sqrt{x})^2}{x} dx = 9 \log x + 12 \sqrt{x} + x + C.$

137. By separation into partial fractions having a common denominator.

Since $\dfrac{f(x) + \phi(x)}{F(x)} dx = \dfrac{f(x)}{F(x)} dx + \dfrac{\phi(x)}{F(x)} dx,$

a fraction may be separated into partial fractions having a common denominator, and thus integrated, if the partial fractions are integrable.

EXAMPLES.

1. $\int \dfrac{a + bx}{1 + x^2}\, dx = \int \dfrac{a\, dx}{1 + x^2} + \int \dfrac{bx\, dx}{1 + x^2}$

$\qquad\qquad = a \tan^{-1} x + \log (1 + x^2)^{\frac{b}{2}} + C.$

2. $\int \sqrt{\dfrac{1 + x}{1 - x}}\, dx = \int \dfrac{1 + x}{\sqrt{1 - x^2}}\, dx = \sin^{-1} x - (1 - x^2)^{\frac{1}{2}} + C.$

3. $\int \dfrac{(x^2 - 1)^{\frac{1}{2}}}{x}\, dx = (x^2 - 1)^{\frac{1}{2}} - \sec^{-1} x + C.'$

4. $\int \dfrac{-x\, dx}{\sqrt{ax - x^2}} = \int \dfrac{a - 2x - a}{2\sqrt{ax - x^2}}\, dx$

$\qquad\qquad = \int \dfrac{a - 2x}{2\sqrt{ax - x^2}}\, dx - \dfrac{a}{2} \int \dfrac{dx}{\sqrt{ax - x^2}}$

$\qquad\qquad = (ax - x^2)^{\frac{1}{2}} - \dfrac{a}{2}\, \mathrm{vers}^{-1} \dfrac{2}{a} x + C.$

5. $\int \dfrac{dx}{x^2 + 4x} = \dfrac{1}{4} \int \dfrac{x + 4 - x}{x^2 + 4x}\, dx = \dfrac{1}{4} \int \dfrac{-x\, dx}{x^2 + 4x} + \dfrac{1}{4} \int \dfrac{x + 4}{x^2 + 4x}\, dx$

$\qquad\qquad = \dfrac{1}{4} \int \dfrac{-dx}{x + 4} + \dfrac{1}{4} \int \dfrac{dx}{x} = \log \left[c \left(\dfrac{x}{x + 4} \right)^{\frac{1}{4}} \right].$

CHAPTER VII.

GENERAL METHODS OF REDUCTION.

BY PARTIAL FRACTIONS.

138. Rational Fractions. Every fraction of the form

$$\frac{ax^m + bx^{m-1} + \cdots lx + k}{a'x^n + b'x^{n-1} + \cdots l'x + k'},$$

in which m and n are positive integers, is called a **rational fraction.**

It is evident that every such fraction can be reduced by division to a series of monomial terms plus a rational fraction whose numerator is of a lower degree than its denominator. Thus,

$$\frac{x^5}{x^2 - x + 1} = x^3 + x^2 - 1 + \frac{1 - x}{x^2 - x + 1}.$$

As the monomial terms can be integrated, we are concerned only with rational fractions whose numerators are of a lower degree than their denominators, and we are to show, —

1°. That every such fraction can be resolved into partial fractions whose denominators are factors of the denominator of the given fraction, and

2°. That these partial fractions can always be integrated.

There will be four cases, according as the factors of the denominator of the given fraction are

1. real and unequal,	3. imaginary and unequal,
2. real and equal,	4. imaginary and equal.

184

139. CASE 1. *The factors real and unequal.*

1°. Let $\dfrac{f(x)}{\phi(x)}$ be the fraction, $\phi(x)$ being resolvable into n real and unequal factors $x-a,\ x-b,\ \cdots x-n$. Then, to every factor $x-a,\ x-b.\ \cdots x-n$, there corresponds a partial fraction

$$\frac{A}{x-a}.\quad \frac{B}{x-b}.\ \cdots\ \frac{N}{x-n}.$$

in which $A.\ B.\ \cdots\ N$ are constants, or

$$\frac{f(x)}{\phi(x)}=\frac{A}{x-a}+\frac{B}{x-b}+\cdots\frac{N}{x-n}.$$

It is required that this equation shall be an identical one, true for all values of x. Reducing the second member to a common denominator, this denominator will be, by hypothesis, $\phi(x)$, and the sum of the numerators will be equal to $f(x)$. This sum will be a polynomial of the $(n-1)$th degree, and since the equation formed by placing it equal to $f(x)$ must be true for all values of x, the coefficients of like powers of x must be separately equal. We shall therefore have n equations of condition from which to find the values of the n constants $A, B, \cdots N$. Hence the resolution can always be effected.

2°. The integration of $\dfrac{f(x)}{\phi(x)}\,dx$ is thus made to depend upon that of a series of fractions of the same form, namely $\dfrac{A\,dx}{x-a}$. But $\displaystyle\int \frac{A\,dx}{x-a}=A\log(x-a)$. Hence the integration is always possible.

EXAMPLES.

1. $\dfrac{5x^3+1}{x^2-3x+2}\,dx=5x+15+\dfrac{35x-29}{x^2-3x+2}$ by division.

The factors of x^2-3x+2 are $x-1$ and $x-2$; hence

$$\frac{35x-29}{x^2-3x+2}=\frac{A}{x-1}+\frac{B}{x-2}=\frac{A(x-2)+B(x-1)}{x^2-3x+2},$$

whence $\quad\quad 35x - 29 = A(x - 2) + B(x - 1)$
$$= (A + B)x - 2A - B.$$

Equating the coefficients of like powers,

$$35 = A + B, \quad 29 = 2A + B;$$

therefore $A = -6$, $B = 41$, and

$$\int \frac{5x^2+1}{x^2-3x+2}dx = \int 5xdx + \int 15dx - \int \frac{6\,dx}{x-1} + \int \frac{41\,dx}{x-2}$$
$$= \tfrac{5}{2}x^2 + 15x - 6\log(x-1) + 41\log(x-2) + C$$
$$= \tfrac{5}{2}x^2 + 15x + \log\frac{(x-2)^{41}}{(x-1)^6} + C.$$

2. $\dfrac{2x^2 - 3x + 5}{x^3 - 7x^2 + 36}dx$.   The roots of $x^3 - 7x^2 + 36 = 0$ are 6, 3, and -2.

Hence $\dfrac{2x^2 - 3x + 5}{x^3 - 7x^2 + 36} = \dfrac{A}{x-6} + \dfrac{B}{x-3} + \dfrac{C}{x+2}$,

and $\quad 2x^2 - 3x + 5 = A(x-3)(x+2) + B(x-6)(x+2)$
$$+ C(x-6)(x-3).$$

Instead of proceeding as in Ex. 1, the values of the constants are readily found by assuming some value for x, since the equation is to be true for all values of x.   Thus, making x equal to -2, 3, and 6, in succession, we find $C = \tfrac{19}{40}$, $B = -\tfrac{14}{15}$, $A = \tfrac{59}{24}$, and

$$\int \frac{2x^2 - 3x + 5}{x^3 - 7x^2 + 36}dx = \log\left[c\,\frac{(x-6)^{\frac{59}{24}}(x+2)^{\frac{19}{40}}}{(x-3)^{\frac{14}{15}}} \right].$$

3. $\int \dfrac{x+5}{x^2-4}dx = \log\sqrt[4]{\dfrac{(x-2)^7}{(x+2)^3}}\,c.$

4. $\int \dfrac{5x+1}{x^2+x-2}dx = \log\left[c(x-1)^2(x+2)^3\right].$

5. $\int \dfrac{x^2-1}{x^2-4}dx = x + \log\left(\dfrac{x-2}{x+2}\right)^{\frac{3}{4}} + C.$

6. $\int \dfrac{3\,dx}{x^2-9} = \log\sqrt{\dfrac{c^2(x-3)}{x+3}}$.

7. $\int \dfrac{a\,dx}{x^2+bx} = \log\left[c\left(\dfrac{x}{x+b}\right)^{\frac{a}{b}}\right]$.

8. $\int \dfrac{2x+1}{x^2-x-6}\,dx = \log\left[c(x-3)^{\frac{7}{5}}(x+2)^{\frac{3}{5}}\right]$.

9. $\int \dfrac{x^2+8x+4}{x^3+x^2-4x-4}\,dx = \log\left[c\,\dfrac{(x+1)(x-2)^2}{(x+2)^2}\right]$.

10. $\int \dfrac{3x^2-1}{x^3-x}\,dx = \log\left[c(x+1)(x-1)x\right] = \log\left[c(x^3-x)\right]$.

(Form 2.)

140. CASE 2. *The factors real and equal.*

1°. Let $\dfrac{f(x)}{\phi(x)}$ be the fraction, $\phi(x)$ being resolvable into n real and equal factors $x-a,\ x-a,\ \cdots$. Then, to such set of n equal factors there corresponds a set of n partial fractions,

$\dfrac{A}{(x-a)^n},\ \dfrac{B}{(x-a)^{n-1}},\ \cdots \dfrac{N}{x-a}$, in which $A,\ B,\ \cdots N$ are constants, or $\dfrac{f(x)}{\phi(x)} = \dfrac{A}{(x-a)^n} + \dfrac{B}{(x-a)^{n-1}} + \cdots \dfrac{N}{x-a}$.

Reducing the second number to a common denominator, this denominator will be equal to $\phi(x)$, and the sum of the numerators will be a polynomial of the $(n-1)$th degree equal to $f(x)$. The latter equation is to be an identical one true for all values of x; hence, equating separately the coefficients of the like powers of x, we have n equations of condition from which to find the values of the n constants $A,\ B,\ \cdots N$. The resolution is, therefore, always possible.

When the factors of $\phi(x)$ are not all equal, the two cases can be combined. Thus

$$\frac{f(x)}{(x-2)^3(x-3)^2(x-4)} = \frac{A}{(x-2)^3} + \frac{B}{(x-2)^2} + \frac{C}{x-2}$$
$$+ \frac{D}{(x-3)^2} + \frac{E}{x-3} + \frac{F}{x-4}.$$

2°. The integration of $\dfrac{f(x)}{\phi(x)}dx$ is thus made to depend upon that of a series of fractions of the form $\dfrac{Adx}{(x-a)^n}$. If $n=1$,

$$\int \frac{Adx}{x-a} = A\log(x-a).$$ If n is other than 1,

$$\int \frac{Adx}{(x-a)^n} = A\int(x-a)^{-n}dx = \frac{A}{1-n}(x-a)^{1-n}.$$

Hence the integration is always possible.

EXAMPLES.   1. $\dfrac{x-1}{(x+1)^2}\,dx.$

Placing

$$\frac{x-1}{(x+1)^2} = \frac{A}{(x+1)^2} + \frac{B}{x+1} = \frac{A+B(x+1)}{(x+1)^2},$$

and equating the numerators, we have $x-1 = A + Bx + B$. Placing the coefficients of like powers equal, we obtain $B=1$, $A=-2$; whence

$$\int \frac{x-1}{(x+1)^2}dx = \int \frac{-2\,dx}{(x+1)^2} + \int \frac{dx}{x+1} = \frac{2}{x+1} + \log(x+1) + C.$$

2. $\dfrac{(x^4+5)dx}{(x-1)^3(x+2)(x+1)}.$

This is a combination of Cases 1 and 2, three only of the factors being equal.   Hence we assume

$$\frac{x^4+5}{(x-1)^3(x+2)(x+1)} = \frac{A}{(x-1)^3} + \frac{B}{(x-1)^2} + \frac{C}{x-1}$$
$$+ \frac{D}{x+2} + \frac{E}{x+1},$$

whence, reducing to a common denominator, and equating the numerators,

$$x^4 + 5 = A(x+2)(x+1) + B(x-1)(x+2)(x+1)$$
$$+ C(x-1)^2(x+2)(x+1)$$
$$+ D(x-1)^3(x+1) + E(x-1)^3(x+2), \qquad (1)$$

$$. \quad =(C+D+E)x^4 +(B+C-2D-E)x^3$$
$$+(A+2B-3C-3E)x^2$$
$$+(3A-B-C+2D+5E)x$$
$$+2A-2B+2C-D-2E. \tag{2}$$

In (1), make $x=-1$, $x=-2$, $x=1$, in succession, and we have directly $E=-\frac{3}{4}$, $D=\frac{7}{9}$, $A=1$. Equating the coefficients of x^4 in (2), $C+D+E=1$, whence, having D and E, $C=\frac{35}{36}$. Equating also the absolute terms,

$$5=2A-2B+2C-D-2E,$$

whence $B=-\frac{1}{6}$. Therefore

$$\int \frac{(x^4+5)dx}{(x-1)^3(x+2)(x+1)} = \int \frac{dx}{(x-1)^3} - \frac{1}{6}\int \frac{dx}{(x-1)^2} + \frac{35}{36}\int \frac{dx}{x-1}$$

$$+\frac{7}{9}\int \frac{dx}{x+2} - \frac{3}{4}\int \frac{dx}{x+1} = -\frac{1}{2(x-1)^2} + \frac{1}{6}\frac{1}{x-1}$$

$$+\frac{35}{36}\log(x-1) + \frac{7}{9}\log(x+2) - \frac{3}{4}\log(x+1) + C.$$

3. $\int \dfrac{3x-1}{(x-3)^2}dx = -\dfrac{8}{x-3} + \log(x-3)^3 + C.$

4. $\int \dfrac{(2+x)dx}{(x-1)^2(x-2)} = \dfrac{3}{x-1} + 4\log\dfrac{x-2}{x-1} + C.$

5. $\int \dfrac{(2x-5)dx}{x^3+5x^2+7x+3} = \dfrac{7}{2(x+1)} + \frac{11}{4}\log\dfrac{x+1}{x+3} + C.$

6. $\int \dfrac{dx}{(x-2)^2(x+3)^2} = -\dfrac{1}{25}\left(\dfrac{1}{x-2}+\dfrac{1}{x+3}\right) + \frac{2}{125}\log\dfrac{x+3}{x-2} + C.$

7. $\int \dfrac{(x+2)dx}{(x-1)^3(x+1)} = \dfrac{1}{4}\left(\dfrac{1}{x-1}-\dfrac{3}{(x-1)^2}\right) + \frac{1}{8}\log\dfrac{x-1}{x+1} + C.$

8. $\int \dfrac{3x-1}{(x-3)^3}dx = \dfrac{5-3x}{(x-3)^2}.$

141. When the factors of $\phi(x)$ are imaginary, the above processes will lead to the logarithms of imaginary quantities.

To avoid such we resolve $\phi(x)$ into quadratic, instead of simple, factors, as follows :

The general form of an imaginary quantity being $a+b\sqrt{-1}$, that of an imaginary factor will be $x-(a+b\sqrt{-1})$. But for every such factor there must be another, $x-(a-b\sqrt{-1})$, since $\phi(x)$ is real. Therefore, for every *pair* of imaginary factors, $\phi(x)$ will have a *quadratic* factor of the form

$$[x-(a+b\sqrt{-1})][x-(a-b\sqrt{-1})]=(x-a)^2+b^2.$$

CASE 3. *The factors imaginary and unequal.*

1°. Let $\dfrac{f(x)}{\phi(x)}$ be the fraction, $\phi(x)$ being resolvable into p unequal quadratic factors $(x-a)^2+b^2$, $(x-c)^2+d^2$, etc. Then, to every such quadratic factor there corresponds a partial fraction $\dfrac{A+Bx}{(x-a)^2+b^2}$, $\dfrac{C+Dx}{(x-c)^2+d^2}$, etc., in which A, B, C, D, etc., are constants, or

$$\frac{f(x)}{\phi(x)}=\frac{A+Bx}{(x-a)^2+b^2}+\frac{C+Dx}{(x-c)^2+d^2}+\cdots\frac{M+Nx}{(x-m)^2+n^2}.$$

For, in reducing the second member to the common denominator $\phi(x)$, any numerator, as $A+Bx$, will be multiplied by $p-1$ factors of the form $(x-a)^2+b^2$, and the sum of the numerators $[=f(x)]$ will therefore be a polynomial of the $[2(p-1)+1]$th degree. We shall therefore have

$$2(p-1)+2=2p$$

equations of condition from which to find the values of the $2p$ constants A, B, $\cdots$ N, and the resolution is always possible.

2°. The integration of $\dfrac{f(x)}{\phi(x)}\,dx$ is thus made to depend upon that of a series of fractions of the form

$$\frac{(A+Bx)dx}{(x-a)^2+b^2}=\frac{(A+Ba)dx}{(x-a)^2+b^2}+\frac{B(x-a)dx}{(x-a)^2+b^2}.$$

But $\quad \int \dfrac{(A+Ba)dx}{(x-a)^2+b^2} = \dfrac{A+Ba}{b}\tan^{-1}\dfrac{x-a}{b}$ [Art. 133 (15')],

and $\quad \int \dfrac{B(x-a)dx}{(x-a)^2+b^2} = \dfrac{B}{2}\log\left[(x-a)^2+b^2\right].$

Hence the integration is always possible.

EXAMPLES. 1. $\dfrac{(x-4)dx}{x^2-4x+5}.$

The factors of x^2-4x+5 are $x-(2\pm\sqrt{-1})$, and their product is $(x-2)^2+1$, or $a=2$, $b=1$ in the form $(x-a)^2+b^2$.

Assuming $\dfrac{x-4}{x^2-4x+5} = \dfrac{A+Bx}{(x-2)^2+1}$, we have $A=-4$, $B=1$.

Hence

$$\int\dfrac{(x-4)dx}{x^2-4x+5} = \dfrac{A+Ba}{b}\tan^{-1}\dfrac{x-a}{b} + \dfrac{B}{2}\log\left[(x-a)^2+b^2\right]$$

$$= -2\tan^{-1}(x-2) + \tfrac{1}{2}\log\left[(x-2)^2+1\right] + C.$$

2. $\dfrac{(x^3+x^2+x+1)dx}{(x-1)^2(x^2+2)}.$

Assume $\dfrac{x^3+x^2+x+1}{(x-1)^2(x^2+2)} = \dfrac{A}{(x-1)^2} + \dfrac{B}{x-1} + \dfrac{C+Dx}{x^2+2},$

whence $A=\tfrac{4}{3}$, $B=\tfrac{10}{9}$, $C=\tfrac{5}{9}$. $D=-\tfrac{1}{9}$. and

$$\int\dfrac{(x^3+x^2+x+1)dx}{(x-1)^2(x^2+2)} = \dfrac{4}{3}\int\dfrac{dx}{(x-1)^2} + \dfrac{10}{9}\int\dfrac{dx}{x-1}$$

$$+ \dfrac{5}{9}\int\dfrac{dx}{x^2+2} - \dfrac{1}{9}\int\dfrac{xdx}{x^2+2}$$

$$= -\dfrac{4}{3}\dfrac{1}{x-1} + \dfrac{10}{9}\log(x-1)$$

$$+ \dfrac{5}{9\sqrt{2}}\tan^{-1}\dfrac{x}{\sqrt{2}} - \dfrac{1}{18}\log(x^2+2) + C.$$

3. $\dfrac{x^2dx}{(x+1)(x-1)(x^2+2)} = \dfrac{Adx}{x+1} + \dfrac{Bdx}{x-1} + \dfrac{(C+Dx)dx}{x^2+2}.$

Then

$$x^2 = A(x-1)(x^2+2) + B(x+1)(x^2+2)$$
$$+(C+Dx)(x+1)(x-1) \qquad (1)$$
$$=(A+B+D)x^3+(B-A+C)x^2$$
$$+(2A+2B-D)x+2B-2A-C. \qquad (2)$$

From (1), when $x = 1$, $x = -1$, in succession, we have $B = \frac{1}{6}$, $A = -\frac{1}{6}$.

From (2), $B - A + C = 1$, or $C = \frac{2}{3}$; and $A + B + D = 0$, or $D = 0$.

Therefore

$$\int \frac{x^2 dx}{(x+1)(x-1)(x^2+2)} = -\frac{1}{6}\int\frac{dx}{x+1} + \frac{1}{6}\int\frac{dx}{x-1} + \frac{2}{3}\int\frac{x}{x^2+2}$$

$$= \frac{1}{6}\log\frac{x-1}{x+1} + \frac{\sqrt{2}}{3}\tan^{-1}\frac{x}{\sqrt{2}} + C.$$

4. $\displaystyle\int \frac{dx}{(x-1)(x^2+2)} = \frac{1}{3}\log(x-1) - \frac{1}{6}\log(x^2+2)$

$$-\frac{1}{3\sqrt{2}}\tan^{-1}\frac{x}{\sqrt{2}} + C.$$

5. $\displaystyle\int \frac{(x^2+2)dx}{x^3+x^2+x+1} = \frac{3}{2}\log(x+1) - \frac{1}{4}\log(x^2+1)$

$$+\frac{1}{2}\tan^{-1}x + C.$$

142. CASE 4. *The factors imaginary and equal.*

1°. Let $\phi(x)$ be the fraction, $\dfrac{f(x)}{\phi(x)}$ being resolvable into p equal quadratic factors $(x-a)^2+b^2$, $(x-a)^2+b^2$, etc. Then, to such set of factors there corresponds a set of p partial fractions,

$$\frac{A+Bx}{[(x-a)^2+b^2]^p}, \quad \frac{C+Dx}{[(x-a)^2+b^2]^{p-1}} \cdots \frac{M+Nx}{(x-a)^2+b^2},$$

in which A, B, $\cdots$ N are constants, or

$$\frac{f(x)}{\phi(x)} = \frac{A+Bx}{[(x-a)^2+b^2]^p} + \frac{C+Dx}{[(x-a)^2+b^2]^{p-1}} + \cdots \frac{M+Nx}{(x-a)^2+b^2}.$$

Reducing the second member to the common denominator $\phi(x)$, the sum of the numerators $[=f(x)]$ will be a polynomial of the $[2(p-1)+1]$th, or $(2p-1)$th, degree. This equation will therefore furnish $2p$ equations of condition, from which the values of the $2p$ constants can always be determined. The resolution is, therefore, always possible.

2°. The integration of $\dfrac{f(x)}{\phi(x)}dx$ is thus made to depend upon that of the general form $\dfrac{(A+Bx)dx}{[(x-a)^2+b^2]^p}$, *in which p is integral.* If $p=1$, the integration has been shown to be possible in Art. 141. If p is other than 1, place $x-a=z$, whence $x=z+a$, $dx=dz$.

Then

$$\int \frac{(A+Bx)dx}{[(x-a)^2+b^2]^p}$$

$$=\int\frac{(A+Bz+Ba)dz}{(z^2+b^2)^p}=\int\frac{Bzdz}{(z^2+b^2)^p}+\int\frac{(A+Ba)dz}{(z^2+b^2)^p}$$

$$=-\frac{B}{2(p-1)(z^2+b^2)^{p-1}}+(A+Ba)\int\frac{dz}{(z^2+b^2)^p};$$

and it will be shown in Art. 147, that the integration of $\dfrac{dz}{(z^2+b^2)^p}$ is always possible when p is integral.

EXAMPLES.

1. $\displaystyle\int\frac{(x^3+x^2+2)dx}{(x^2+2)^2}=\int\frac{(A+Bx)dx}{(x^2+2)^2}+\int\frac{(C+Dx)dx}{x^2+2}$,

whence $x^3+x^2+2=A+Bx+(C+Dx)(x^2+2)$. from which we find $A=0$, $B=-2$, $C=D=1$. Hence

$$\int\frac{(x^3+x^2+2)dx}{(x^2+2)^2}=\int\frac{-2xdx}{(x^2+2)^2}+\int\frac{dx}{x^2+2}+\int\frac{xdx}{x^2+2}$$

$$=\frac{1}{x^2+2}+\frac{1}{\sqrt{2}}\tan^{-1}\frac{x}{\sqrt{2}}+\tfrac{1}{2}\log(x^2+2)+C.$$

2. $\int \frac{(x^4 + 2x^3 - 2x^2 - 2x + 5)dx}{(x^2+1)^2(x-2)} = \frac{A+Bx}{(x^2+1)^2} + \frac{C+Dx}{x^2+1} + \frac{E}{x-2}$,

whence $A = -4$, $B = 0$, $C = 2$, $D = 0$, $E = 1$, and we have

$$\int \frac{-4\,dx}{(x^2+1)^2} + \int \frac{2\,dx}{x^2+1} + \int \frac{dx}{x-2}$$

$$= 2\tan^{-1}x + \log(x-2) - 4\int \frac{dx}{(x^2+1)^2}.$$

3. $\int \frac{(x^2 - x + 1)dx}{(x+1)(x^2+1)} = \log \frac{(x+1)^{\frac{3}{2}}}{(x^2+1)^{\frac{1}{4}}} - \frac{1}{2}\tan^{-1}x + C.$

4. $\int \frac{x^7 + x^5 + x^3 + x}{(x^2+2)^2(x^2+3)^2}dx = \frac{5}{2(x^2+2)} + \frac{10}{x^2+3}$

$$+ \tfrac{19}{2}\log(x^2+2) - 9\log(x^2+3) + C.$$

BY RATIONALIZATION.

Since rational algebraic polynomials and rational fractions can always be integrated, an irrational differential may be integrated if it can be rationalized. The rationalization is effected by substituting for the variable of the given differential a new variable of which it is a function. Of these substitutions the following are the most important:

143. When the only function of x affected with fractional exponents is a linear one, in which case it will be either of the form $x^{\frac{p}{q}}$ or $(ax + b)^{\frac{p}{q}}$, assume $x = z^n$ or $ax + b = z^n$, n being the least common multiple of the denominators of the fractional exponents. For, if $x = z^n$ or $ax + b = z^n$, the values of x, $ax + b$, dx, and the surds of the given differential will be rational functions of z.

EXAMPLES.   1. $\int \frac{x^{\frac{2}{3}} - x^{\frac{1}{4}}}{x^{\frac{1}{2}}} dx.$

Here $n = 12$, and $x = z^{12}$.

Hence $x^{\frac{2}{3}} = z^8$, $x^{\frac{1}{4}} = z^3$, $x^{\frac{1}{2}} = z^6$, $dx = 12z^{11}dz$;

$$\therefore \int \frac{x^{\frac{2}{3}} - x^{\frac{1}{4}}}{x^{\frac{1}{2}}} dx = \int \frac{z^8 - z^3}{z^6} 12 z^{11} dz = 12 \int (z^{13} - z^8) dz$$

$$= \tfrac{6}{7} z^{14} - \tfrac{4}{3} z^9 + C = \tfrac{6}{7} x^{\frac{7}{6}} - \tfrac{4}{3} x^{\frac{3}{4}} + C.$$

2. $\displaystyle \int \frac{x^{\frac{3}{2}} - x^{\frac{1}{4}}}{6 x^{\frac{1}{4}}} dx = \tfrac{2}{3} (\tfrac{1}{9} x^{\frac{9}{4}} - \tfrac{3}{13} x^{\frac{13}{12}}) + C.$

3. $\displaystyle \int \frac{dx}{x^{\frac{1}{2}} - x^{\frac{1}{3}}} = 2 x^{\frac{1}{2}} + 3 x^{\frac{1}{3}} + 6 x^{\frac{1}{6}} + 6 \log(x^{\frac{1}{6}} - 1) + C.$

4. $\displaystyle \int \frac{(2 - x)^{\frac{1}{2}} dx}{3 - x} = \int \frac{(2 - x)^{\frac{1}{2}} dx}{1 + (2 - x)}.$

Assume $2 - x = z^2$.

Then, $(2 - x)^{\frac{1}{2}} = z$, $dx = -2 z dz$,

and $\displaystyle \int \frac{(2 - x)^{\frac{1}{2}} dx}{3 - x} = \int \frac{-2 z^2 dz}{1 + z^2} = -2 \int \left(1 - \frac{1}{1 + z^2}\right) dz$

$$= -2 (z + \cot^{-1} z) + C$$

$$= -2 (\cot^{-1} \sqrt{2 - x} + \sqrt{2 - x}) + C.$$

5. $\displaystyle \int \frac{y dy}{(2 r - y)^{\frac{1}{2}}} = -\tfrac{2}{3} (4 r + y)(2 r - y)^{\frac{1}{2}} + C.$

6. $\displaystyle \int \frac{dx}{(x + 1)^{\frac{1}{3}} - (x + 1)^{\frac{1}{2}}}$

$$= 3 [(x + 1)^{\frac{1}{3}} + 2 (x + 1)^{\frac{1}{6}} + 2 \log((x + 1)^{\frac{1}{6}} - 1)] + C.$$

7. $\displaystyle \int x^2 (1 + x)^{\frac{1}{2}} dx = 2 (1 + x)^{\frac{3}{2}} [(1 + x)^2 - \tfrac{2}{5} (1 + x) + \tfrac{1}{7}] + C.$

144. When the only surd of the given differential is of the form $\sqrt{a + bx \pm x^2}$, rationalization is effected as follows:

I. *When the sign of x^2 is positive*, place $\sqrt{a + bx + x^2} = z - x$.

Then $a + bx = z^2 - 2 zx$;

whence $x = \dfrac{z^2 - a}{b + 2 z}$, $dx = \dfrac{2 (z^2 + bz + a) dz}{(b + 2 z)^2}$, and $\sqrt{a + bx + x^2} = z - x = \dfrac{z^2 + bz + a}{b + 2z}.$

The given differential will then be a rational function of z, since x, dx, and $\sqrt{a + bx + x^2}$ are rational functions of z.

II. *When the sign of x^2 is negative*, place

$$\sqrt{a + bx - x^2} = \sqrt{(x - a)(\beta - x)} = (x - a)z,$$

in which a and β are the roots of $x^2 - bx - a = 0$.

Then $\beta - x = (x - a)z^2;$

whence $x = \dfrac{az^2 + \beta}{z^2 + 1}, \quad dx = \dfrac{2(a - \beta)zdz}{(z^2 + 1)^2},$

and $\sqrt{a + bx - x^2} = (x - a)z = \dfrac{(\beta - a)z}{z^2 + 1}.$

The given differential will then be a rational function of z, since x, dx, and $\sqrt{a + bx - x^2}$ are rational functions of z.

EXAMPLES.   1. $\displaystyle\int \dfrac{dx}{\sqrt{1 + x + x^2}}.$

Assume $\sqrt{1 + x + x^2} = z - x.$

Then $x = \dfrac{z^2 - 1}{1 + 2z};$

whence $dx = \dfrac{2(z^2 + z + 1)dz}{(1 + 2z)^2},$

$$\sqrt{1 + x + x^2} = z - x = \dfrac{z^2 + z + 1}{1 + 2z}.$$

Hence $\displaystyle\int \dfrac{dx}{\sqrt{1 + x + x^2}} = \int \dfrac{2\,dz}{1 + 2z} = \log(1 + 2z) + C$

$$= \log(1 + 2x + 2\sqrt{1 + x + x^2}) + C.$$

2. $\displaystyle\int \dfrac{dx}{x^2 - x - 1} = \log(2\sqrt{x^2 - x - 1} + 2x - 1) + C.$

3. $\displaystyle\int \dfrac{(2x + x^2)^{\frac{1}{2}}dx}{x^2} = \log(x + 1 + \sqrt{2x + x^2}) - \dfrac{4}{x + \sqrt{2x + x^2}} + C.$

4. $\int \dfrac{dx}{\sqrt{1+2x+x^2}} = \int \dfrac{dx}{1+x} = \log(1+x) + C.$

Or, by the above method,

$$\int \dfrac{dx}{\sqrt{1+2x+x^2}} = \log(1+x+\sqrt{1+2x+x^2}) + C''$$

$$= \log 2(1+x) + C'.$$

Prove that $C' = C - \log 2.$

5. $\int \dfrac{dx}{\sqrt{x^2+x}} = \log(\tfrac{1}{2} + x + \sqrt{x^2+x}) + C.$

6. $\int \dfrac{dx}{\sqrt{2+x-x^2}}.$

The roots of $x^2 - x - 2 = 0$ are 2 and -1.

Hence $x^2 - x - 2 = (x-2)(x+1),$

and $\sqrt{2+x-x^2} = \sqrt{(x+1)(2-x)} = (x+1)z.$

Squaring, we find $x = \dfrac{2-z^2}{z^2+1};$ whence

$$dx = \dfrac{-6z\,dz}{(z^2+1)^2}, \quad \sqrt{2+x-x^2} = (x+1)z = \dfrac{3z}{z^2+1}.$$

Hence $\int \dfrac{dx}{\sqrt{2+x-x^2}} = \int -\dfrac{z\,dz}{z^2+1} = 2 \cot^{-1} z + C$

$$= 2 \cot^{-1} \sqrt{\dfrac{2-x}{x+1}} + C.$$

7. $\int \dfrac{dx}{\sqrt{2-x-x^2}} = 2 \cot^{-1} \sqrt{\dfrac{1-x}{x+2}} + C.$

8. $\int \dfrac{dx}{\sqrt{4x-3-x^2}} = 2 \cot^{-1} \sqrt{\dfrac{1-x}{x-3}} + C.$

9. $\int \dfrac{dx}{\sqrt{2-2x-x^2}} = 2 \cot^{-1} \sqrt{\dfrac{\sqrt{3}-1-x}{\sqrt{3}+1+x}} + C.$

145. Binomial Differentials. Every binomial differential may be reduced to the form

$$x^m(a + bx^n)^p dx,$$

in which p may be any fraction, but m and n are integral and n positive.

For, let $x^h(ax^k + bx^t)^p dx$ be the binomial, and let $k < t$. Then

$$x^h(ax^k + bx^t)^p dx = x^h x^{pk}\left(a + b\frac{x^t}{x^k}\right)^p dx = x^{h+pk}(a + bx^{t-k})^p dx;$$

in which $t - k$ may be fractional, but is positive, and $h + pk$ is fractional or integral, positive or negative. That is, the binomial is of the form

$$x^{\pm\frac{c}{d}}(a + bx^{+\frac{e}{f}})^p dx.$$

Put $x = z^{df}$, and this becomes

$$z^{\pm cf}(a + bz^{+de})^p df z^{df-1} dz = df z^{\pm cf + df - 1}(a + bz^{+de})^p dz,$$

in which $\pm cf + df - 1$ and de are integral, and the latter positive.

Hence writing m for the former and n for the latter, we have $df z^m(a + bz^n)^p dz$, which is of the required form. As p may be fractional, represent it by $\frac{r}{s}$.

We are now to show that a binomial differential of the form $x^m(a + bx^n)^{\frac{r}{s}} dx$, in which m and n are integral and n positive, may be rationalized, and therefore integrated:

I. *When* $\dfrac{m+1}{n}$ *is a whole number or zero, by assuming* $a + bx^n = z^s$.

II. *When* $\dfrac{m+1}{n} + \dfrac{r}{s}$ *is a whole number or zero, by assuming* $a + bx^n = z^s x^n$.

To prove that the rationalization is effected when the above conditions are satisfied:

I. Let $a + bx^n = z^s$.

Then $x = \left(\dfrac{z^s - a}{b}\right)^{\frac{1}{n}}$, $x^m = \left(\dfrac{z^s - a}{b}\right)^{\frac{m}{n}}$, $dx = \dfrac{s}{nb}\left(\dfrac{z^s - a}{b}\right)^{\frac{1}{n} - 1} z^{s-1} dz$,

and $\qquad (a + bx^n)^{\frac{r}{s}} = z^r$.

Hence

$$x^m (a + bx^n)^{\frac{r}{s}} dx = \left(\dfrac{z^s - a}{b}\right)^{\frac{m}{n}} \cdot z^r \cdot \dfrac{s}{nb}\left(\dfrac{z^s - a}{b}\right)^{\frac{1}{n} - 1} z^{s-1} dz$$

$$= \dfrac{s}{nb} z^{r+s-1}\left(\dfrac{z^s - a}{b}\right)^{\frac{m+1}{n} - 1} dz,$$

which is rational when $\dfrac{m + 1}{n}$ is a whole number or zero.

II. Let $a + bx^n = z^s x^n$.

Then $\qquad x = \left(\dfrac{a}{z^s - b}\right)^{\frac{1}{n}}$, $x^m = \left(\dfrac{a}{z^s - b}\right)^{\frac{m}{n}}$,

$$dx = -\dfrac{as}{n}\left(\dfrac{a}{z^s - b}\right)^{\frac{1}{n} - 1} \dfrac{z^{s-1}}{(z^s - b)^2} dz,$$

and $\qquad (a + bx^n)^{\frac{r}{s}} = z^r x^{\frac{nr}{s}} = z^r \left(\dfrac{a}{z^s - b}\right)^{\frac{r}{s}}$.

Hence $\quad x^m (a + bx^n)^{\frac{r}{s}} dx$

$$= -\left(\dfrac{a}{z^s - b}\right)^{\frac{m}{n}} \cdot z^r \left(\dfrac{a}{z^s - b}\right)^{\frac{r}{s}} \cdot \dfrac{as}{n}\left(\dfrac{a}{z^s - b}\right)^{\frac{1}{n} - 1} \dfrac{z^{s-1}}{(z^s - b)^2} dz$$

$$= -\dfrac{s}{n} a^{\frac{m+1}{n} + \frac{r}{s}} z^{r+s-1}\left(\dfrac{1}{z^s - b}\right)^{\frac{m+1}{n} + \frac{r}{s} + 1} dz,$$

which is rational when $\dfrac{m + 1}{n} + \dfrac{r}{s}$ is a whole number or zero. When $\dfrac{r}{s}$ is a positive integer, the factor $(a + bx^n)^{\frac{r}{s}}$ may be expanded and integrated directly.

EXAMPLES. 1. $\displaystyle\int \dfrac{x^3 dx}{(a + bx^2)^{\frac{3}{2}}} = \int x^3 (a + bx^2)^{-\frac{3}{2}} dx$.

Here $\qquad \dfrac{m + 1}{n} = 2$.

Assume therefore

$$a + bx^2 = z^2; \quad \text{whence} \quad (a + bx^2)^{\frac{3}{2}} = z^3,$$

$$x^3 = \left(\frac{z^2 - a}{b}\right)^{\frac{3}{2}}, \quad dx = \frac{zdz}{b^{\frac{1}{2}}(z^2 - a)^{\frac{1}{2}}}.$$

Hence

$$\int \frac{x^3 dx}{(a + bx^2)^{\frac{3}{2}}} = \int \left(\frac{z^2 - a}{b}\right)^{\frac{3}{2}} \cdot \frac{zdz}{b^{\frac{1}{2}}(z^2 - a)^{\frac{1}{2}}} \cdot \frac{1}{z^3}$$

$$= \frac{1}{b^2}\int \left(1 - \frac{a}{z^2}\right)dz = \frac{1}{b^2}\left(z + \frac{a}{z}\right) + C$$

$$= \frac{1}{b^2}\frac{2a + bx^2}{\sqrt{a + bx^2}} + C.$$

2. $\displaystyle\int \frac{x^3 dx}{(a^2 + x^2)^{\frac{1}{2}}} = \frac{(a^2 + x^2)^{\frac{1}{2}}}{3}(x^2 - 2a^2) + C.$

3. $\displaystyle\int x(1 + x)^{\frac{3}{2}}dx = \frac{2}{35}(1 + x)^{\frac{5}{2}}(5x - 2) + C.$

4. $\displaystyle\int \frac{x^2 dx}{(a + bx^2)^{\frac{5}{2}}} = \int x^2(a + bx^2)^{-\frac{5}{2}}dx.$

Here $\quad \dfrac{m + 1}{n} + \dfrac{r}{s} = -1.$

Assume therefore

$$a + bx^2 = z^2 x^2; \quad \text{whence} \quad x^2 = \frac{a}{z^2 - b},$$

$$(a + bx^2)^{\frac{5}{2}} = z^5\left(\frac{a}{z^2 - b}\right)^{\frac{5}{2}}, \quad dx = -\frac{a^{\frac{1}{2}}zdz}{(z^2 - b)^{\frac{3}{2}}}.$$

Hence

$$\int \frac{x^2 dx}{(a + bx^2)^{\frac{5}{2}}} = -\int \frac{a}{z^2 - b} \cdot \frac{a^{\frac{1}{2}}zdz}{(z^2 - b)^{\frac{3}{2}}} \cdot \left(\frac{z^2 - b}{a}\right)^{\frac{5}{2}}\frac{1}{z^5}$$

$$= -\int \frac{dz}{az^4} = \frac{a}{3z^3} + C = \frac{a}{3}\frac{x^3}{(a + bx^2)^{\frac{3}{2}}} + C.$$

5. $\displaystyle\int \frac{dx}{x(a^2 - x^2)^{\frac{1}{2}}} = \frac{1}{a}\log\frac{x}{\sqrt{a^2 - x^2} + a} + C.$

6. $\int \dfrac{dx}{x^2(a + bx^2)^{\frac{3}{2}}} = -\dfrac{1}{a^2}\dfrac{a + 2bx^2}{x(a + bx^2)^{\frac{1}{2}}} + C.$

7. $\int \dfrac{dx}{x^4\sqrt{1 - 2x^2}} = -\dfrac{1 + 4x^2}{3x^3}\sqrt{1 - 2x^2} + C.$

8. $\int x^3(1 + 2x^2)^{\frac{3}{2}}dx = (1 + 2x^2)^{\frac{5}{2}}\dfrac{5x^2 - 1}{70} + C.$

BY PARTS.

146. Let u and v be any functions of x. Then

$$d(uv) = udv + vdu.$$

Transposing and integrating,

$$\int udv = uv - \int vdu.$$

This formula is known as the formula for integration by parts. It evidently makes the integration of udv to depend upon that of vdu. To apply it, the given differential must be resolved into factors u and dv such that dv and vdu shall be integrable. The following are the most important applications of this formula.

147. Binomial differentials. Formulæ of reduction.

It has been shown that every binomial differential may be reduced to the form $x^m(a + bx^n)^p dx$, in which p is any fraction, but m and n are integral and n positive.

I. Let $u = x^{m-n+1}$, $dv = (a + bx^n)^p x^{n-1}dx$.

Then $du = (m - n + 1)x^{m-n}dx$, $v = \dfrac{(a + bx^n)^{p+1}}{nb(p + 1)}.$

Substituting these in $\int udv = uv - \int vdu$,

$$\int x^m(a + bx^n)^p dx = \dfrac{x^{m-n+1}(a + bx^n)^{p+1}}{nb(p + 1)}$$
$$-\dfrac{m - n + 1}{nb(p + 1)}\int x^{m-n}(a + bx^n)^{p+1}dx.$$

But

$$\int x^{m-n}(a + bx^n)^{p-1}dx = \int x^{m-n}(a + bx^n)^p(a + bx^n)dx$$

$$= a\int x^{m-n}(a+bx^n)^p dx + b\int x^m(a+bx^n)^p dx.$$

Hence

$$\int x^m(a + bx^n)^p dx = \frac{x^{m-n+1}(a + bx^n)^{p+1}}{nb(p + 1)}$$

$$- a\frac{m - n + 1}{nb(p + 1)}\int x^{m-n}(a+bx^n)^p dx - \frac{m-n+1}{n(p+1)}\int x^m(a+bx^n)^p dx,$$

or, solving for $\int x^m(a + bx^n)^p dx$,

$$\int x^m(a + bx^n)^p dx$$

$$= \frac{x^{m-n+1}(a + bx^n)^{p-1} - a(m - n + 1)\int x^{m-n}(a + bx^n)^p dx}{b(np + m + 1)}, \quad (A)$$

a formula which makes the integration of the given binomial to depend upon that of another in which the exponent of the variable without the parenthesis is diminished by that of the variable within.

ILLUSTRATION. $\int\frac{x^3 dx}{\sqrt{1 - x^2}} = \int x^3(1 - x^2)^{-\frac{1}{2}}dx$. We apply (A) to this differential because its integration would thereby be made to depend upon that of $x(1 - x^2)^{-\frac{1}{2}}dx$, which comes under Form 1. Substituting therefore in (A) $m = 3, n = 2, p = -\frac{1}{2}, a = 1, b = -1$, we have

$$\int x^3(1 - x^2)^{-\frac{1}{2}}dx = \frac{x^2(1 - x^2)^{\frac{1}{2}} - 2\int x(1 - x^2)^{-\frac{1}{2}}dx}{-3}$$

$$= -\frac{1}{3}x^2(1 - x^2)^{\frac{1}{2}} + \frac{2}{3}\int x(1 - x^2)^{-\frac{1}{2}}dx$$

$$= -\frac{1}{3}x^2(1 - x^2)^{\frac{1}{2}} - \frac{2}{3}(1 - x^2)^{\frac{1}{2}} + C.$$

If $np + m + 1 = 0$, the formula fails; but in this case

$$\frac{m+1}{n} + p = 0.$$

and the differential may be rationalized and integrated by Art. 145.

II. $\int x^m (a + bx^n)^p dx$

$$= \int x^m (a + bx^n)^{p-1}(a + bx^n) dx$$

$$= a \int x^m (a + bx^n)^{p-1} dx + b \int x^{m+n}(a + bx^n)^{p-1} dx. \qquad (1)$$

Applying (A) to the last integral of (1), we obtain

$$\int x^{m+n}(a + bx^n)^{p-1} dx$$

$$= \frac{x^{m+1}(a + bx^n)^p - a(m+1) \int x^m (a + bx^n)^{p-1} dx}{b(np + m + 1)},$$

which, substituted in (1), gives

$$\int x^m (a + bx^n)^p dx$$

$$= \frac{x^{m+1}(a + bx^n)^p + anp \int x^m (a + bx^n)^{p-1} dx}{np + m + 1}. \qquad (B)$$

a formula which makes the integration of the given binomial to depend upon that of another in which the exponent of the parenthesis is diminished by 1.

ILLUSTRATION. $\int (a^2 + x^2)^{\frac{1}{2}} dx$. The application of (B) to this differential makes the integration depend upon that of $\dfrac{dx}{\sqrt{a^2 + x^2}}$, which can be rationalized and integrated by Art. 144.

Substituting, therefore, in (B) $m = 0$, $n = 2$, $p = \frac{1}{2}$, $a = a^2$, $b = 1$, we have

$$\int (a^2 + x^2)^{\frac{1}{2}}\, dx = \frac{x(a^2 + x^2)^{\frac{1}{2}} + a^2 \int \dfrac{dx}{\sqrt{a^2 + x^2}}}{2}.$$

Writing $\sqrt{a^2 + x^2} = z - x$, we find

$$\int \frac{dx}{\sqrt{a^2 + x^2}} = \int \frac{dz}{z} = \log z + C = \log (x + \sqrt{a^2 + x^2}) + C.$$

Hence

$$\int (a^2 + x^2)^{\frac{1}{2}}\, dx = \tfrac{1}{2} x(a^2 + x^2)^{\frac{1}{2}} + \frac{a^2}{2} \log (x + \sqrt{a^2 + x^2}) + C.$$

If $np + m + 1 = 0$, the formula fails, but Art. 145 applies as before.

III. In (A) let $m = m + n$. Then

$$\int x^{m+n}(a + bx^n)^p dx$$

$$= \frac{x^{m+1}(a + bx^n)^{p+1} - a(m + 1) \int x^m(a + bx^n)^p dx}{b(np + m + n + 1)},$$

whence

$$\int x^m(a + bx^n)^p dx$$

$$= \frac{x^{m+1}(a + bx^n)^{p+1} - b(np + m + n + 1) \int x^{m+n}(a + bx^n)^p dx}{a(m + 1)}, \quad (C)$$

a formula which makes the integration of the given binomial to depend upon that of another in which the exponent of the variable without the parenthesis is increased by that of the variable within.

ILLUSTRATION. $\displaystyle \int \frac{dx}{x^3(x^2 - 1)^{\frac{1}{2}}} = \int x^{-3}(x^2 - 1)^{-\frac{1}{2}} dx$. By applying (C) the integration is made to depend upon that of $\dfrac{dx}{x\sqrt{x^2 - 1}}$, which is a known form. Hence, making $m = -3$, $n = 2$, $p = -\frac{1}{2}$, $a = -1$, $b = 1$, in (C), we have

$$\int \frac{dx}{x^3(x^2-1)^{\frac{1}{2}}} = \frac{x^{-2}(x^2-1)^{\frac{1}{2}} + \int x^{-1}(x^2-1)^{-\frac{1}{2}}dx}{2}$$

$$= \frac{(x^2-1)^{\frac{1}{2}}}{2\,x^2} + \tfrac{1}{2}\sec^{-1}x + C.$$

If $m = -1$, the formula fails; but in this case $\dfrac{m+1}{n} = 0$, and Art. 145 applies.

IV. In (B) let $p = p + 1$. Then

$$\int x^m(a+bx^n)^{p+1}dx$$

$$= \frac{x^{m+1}(a+bx^n)^{p+1} + an(p+1)\int x^m(a+bx^n)^p dx}{np+n+m+1},$$

whence

$$\int x^m(a+bx^n)^p dx$$

$$= \frac{-x^{m+1}(a+bx^n)^{p+1} + (np+n+m+1)\int x^m(a+bx^n)^{p+1}dx}{an(p+1)}, \quad (D)$$

a formula which makes the integration of the given binomial to depend upon that of another in which the exponent of the parenthesis is increased by 1.

ILLUSTRATION. $\displaystyle \int \frac{dx}{(1+x^2)^3} = \int (1+x^2)^{-3}dx$. By applying (D) *twice*, we see the integration will be made to depend upon that of $\dfrac{dx}{1+x^2}$, which is a known form. Hence, making $m = 0$, $n = 1$, $p = -3$, $a = b = 1$, in (D), we have

$$\int \frac{dx}{(1+x^2)^3} = \frac{-x(1+x^2)^{-2} - 3\int (1+x^2)^{-2}dx}{-4}$$

$$= \frac{x}{4(1+x^2)^2} + \frac{3}{4}\int (1+x^2)^{-2}dx.$$

Applying (D) to the last integral, we have $m = 0$, $n = 2$, $p = -2$, $a = b = 1$, and

$$\int (1 + x^2)^{-2} dx = \frac{-x(1+x^2)^{-1} - \int (1+x^2)^{-1} dx}{-2}$$

$$= \frac{x}{2(1+x^2)} + \tfrac{1}{2} \tan^{-1} x + C.$$

Hence

$$\int \frac{dx}{(1+x^2)^3} = \frac{x}{4(1+x^2)^2} + \frac{3x}{8(1+x^2)} + \tfrac{3}{8} \tan^{-1} x + C.$$

EXAMPLES.

1. $\displaystyle \int \frac{x^2 dx}{(a^2 - x^2)^{\frac{1}{2}}} = -\frac{x}{2}(a^2 - x^2)^{\frac{1}{2}} + \frac{a^2}{2} \sin^{-1} \frac{x}{a} + C.$

2. $\displaystyle \int \frac{x^4 dx}{(a^2 - x^2)^{\frac{1}{2}}} = -\frac{x}{4}\left(x^2 + \frac{3a^2}{2}\right)\sqrt{a^2 - x^2} + \frac{3a^4}{8} \sin^{-1} \frac{x}{a} + C.$

3. $\displaystyle \int \frac{x^5 dx}{(a^2 - x^2)^{\frac{1}{2}}} = -\left(\frac{x^4}{5} + \frac{4a^2 x^2}{15} + \frac{8a^4}{15}\right)\sqrt{a^2 - x^2} + C.$

4. $\displaystyle \int \frac{x^3 dx}{\sqrt{a^2 + x^2}} = \tfrac{1}{3}(x^2 - 2a^2)\sqrt{a^2 + x^2} + C.$

5. $\displaystyle \int \frac{x^3 dx}{(a^2 + x^2)^{\frac{3}{2}}} = \frac{x^2 + 2a^2}{(a^2 + x^2)^{\frac{1}{2}}} + C.$

6. $\displaystyle \int \frac{x^2 dx}{(a^2 + x^2)^{\frac{1}{2}}} = \frac{x}{2}\sqrt{a^2 + x^2} - \tfrac{1}{2} a^2 \log (x + \sqrt{a^2 + x^2}) + C.$

7. $\displaystyle \int (a^2 - x^2)^{\frac{3}{2}} dx = \frac{x}{4}(a^2 - x^2)^{\frac{3}{2}} + \frac{3a^2 x}{8}(a^2 - x^2)^{\frac{1}{2}}$

$$+ \frac{3a^4}{8} \sin^{-1} \frac{x}{a} + C.$$

Apply (B) twice.

8. $\int x^2 (1 - x^2)^{\frac{1}{2}} dx = -\frac{x(1 - x^2)^{\frac{3}{2}}}{4} + \frac{x(1 - x^2)^{\frac{1}{2}}}{8} + \frac{1}{8} \sin^{-1} x + C.$

Apply (A) and (B) in succession.

9. $\int \frac{x^2 dx}{\sqrt{2ax - x^2}} = \int x^{\frac{3}{2}} (2a - x)^{-\frac{1}{2}} dx$

$= -\frac{x + 3a}{2} (2ax - x^2)^{\frac{1}{2}} + \frac{3a^2}{2} \mathrm{vers}^{-1} \frac{x}{a} + C.$

Apply (A) twice.

10. $\int \frac{dx}{x^2 (a^2 - x^2)^{\frac{1}{2}}} = -\frac{\sqrt{a^2 - x^2}}{a^2 x} + C.$

Apply (C).

11. $\int \frac{dx}{x^3 (1 - x^2)^{\frac{1}{2}}} = -\frac{\sqrt{1 - x^2}}{2 x^2} + \frac{1}{2} \log \frac{x}{\sqrt{1 - x^2} + 1} + C.$

12. $\int \frac{dx}{(a^2 + x^2)^2} = \frac{x}{2a^2(a^2 + x^2)} + \frac{1}{2a^3} \tan^{-1} \frac{x}{a} + C.$

13. Show that (A) will reduce the following to known forms:

$\frac{x^m dx}{\sqrt{a^2 - x^2}}$, if m is even and positive; also if m is odd and positive.

$\frac{x^m dx}{\sqrt{a^2 + x^2}}$, if m is even and positive.

$x^m (a^2 \pm x^2)^{\pm \frac{p}{q}}$, if m is odd and positive.

What if m is odd and negative?

14. $\int (r^2 - x^2)^{\frac{1}{2}} dx = \frac{1}{2} x (r^2 - x^2)^{\frac{1}{2}} + \frac{1}{2} r^2 \sin^{-1} \frac{x}{r} + C.$

15. $\int \frac{y^3 dy}{\sqrt{2ry - y^2}} = -\frac{2y^2 + 5r(y + 3r)}{6} \sqrt{2ry - y^2}$

$+ \frac{5}{2} r^3 \mathrm{vers}^{-1} \frac{y}{r} + C.$

148. **Logarithmic differentials of the form** $x^m (\log x)^n \, dx$ may be integrated by parts when n is a positive integer, by placing $x^m dx = dv$, $(\log x)^n = u$, in the formula

$$\int u\,dv = uv - \int v\,du \,;$$

every application of the formula reducing the exponent of the logarithm by unity and thus finally making the integration depend upon $\int x^m dx$.

EXAMPLES. 1. $\int x^2 (\log x)^2 dx$.

Let $x^2 dx = dv$, $(\log x)^2 = u$. Then $v = \dfrac{x^3}{3}$, $du = 2 \log x \dfrac{dx}{x}$, and

$$\int u\,dv = (\log x)^2 \frac{x^3}{3} - \frac{2}{3} \int x^2 \log x\,dx.$$

Placing $x^2 dx = dv$, $u = \log x$, whence $v = \dfrac{x^3}{3}$, $du = \dfrac{dx}{x}$,

$$\int u\,dv = \frac{x^3}{3} \log x - \frac{1}{3} \int x^2 dx = \frac{x^3}{3} \log x - \frac{x^3}{9} + C.$$

Hence $\int x^2 (\log x)^2 dx = \dfrac{x^3}{3} \left[(\log x)^2 - \tfrac{2}{3} \log x + \tfrac{2}{9} \right] + C$.

2. $\int \log x\,dx = x (\log x - 1) + C$.

3. $\int x^3 (\log x)^2 dx = \dfrac{x^4}{4} \left[(\log x)^2 - \tfrac{1}{2} \log x + \tfrac{1}{8} \right] + C$.

149. **Exponential differentials of the form** $x^n e^{ax} \, dx$ may be integrated by parts when n is a positive integer, by placing $x^n = u$, $e^{ax} dx = dv$, in $\int u\,dv = uv - \int v\,du$, every application of the formula reducing the exponent of x^n by unity, and thus finally making the integration depend upon $\int e^{ax} dx$.

EXAMPLES. 1. $\int x^2 e^{ax} x dx.$

Let $e^{ax} dx = dv$, $x^2 = u$; then $v = \dfrac{e^{ax}}{a}$, $du = 2\, x dx$, and

$$\int u dv = \frac{x^2 e^{ax}}{a} - \frac{2}{a} \int e^{ax} x dx.$$

Placing $e^{ax} dx = dv$, $x = u$, whence $v = \dfrac{e^{ax}}{a}$, $du = dx$,

$$\int e^{ax} x dx = \frac{x e^{ax}}{a} - \frac{1}{a} \int e^{ax} dx = \frac{x e^{ax}}{a} - \frac{e^{ax}}{a^2} + C.$$

Hence $\int x^2 e^{ax} dx = \dfrac{e^{ax}}{a}\left(x^2 - \dfrac{2\,x}{a} + \dfrac{2}{a^2}\right) + C.$

2. $\int x^3 e^{ax} dx = e^{ax}\left(\dfrac{x^3}{a} - \dfrac{3\,x^2}{a^2} + \dfrac{6\,x}{a^3} - \dfrac{6}{a^4}\right) + C.$

150. Trigonometric Differentials. By simple transformations, some of which are indicated in the following examples, these may often be reduced to known forms. Otherwise resort must be had to integration by parts.

I. $\sin^n x dx$ and $\cos^n x dx.$

(a) *When n is an odd integer*, we may write

$$\sin^n x dx = (1 - \cos^2 x)^{\frac{n-1}{2}} \sin x dx,$$

and $\qquad \cos^n x dx = (1 - \sin^2 x)^{\frac{n-1}{2}} \cos x dx.$

1. $\int \sin^3 x dx = \int (1 - \cos^2 x)\sin x dx = -\cos x + \tfrac{1}{3}\cos^3 x + C.$

2. $\int \cos^5 x dx = \int (1 - \sin^2 x)^2 \cos x dx$
$$= \sin x - \tfrac{2}{3}\sin^3 x + \tfrac{1}{5}\sin^5 x + C.$$

3. $\int \cos^3 x dx = \sin x - \tfrac{1}{3}\sin^3 x + C.$

(b) *When $n = 2$,* since

$$2\sin^2 x = 1 - \cos 2x, \text{ and } 2\cos^2 x = 1 + \cos 2x,$$

4. $\int \sin^2 x dx = \int (\frac{1}{2} - \frac{1}{2}\cos 2x) dx = \frac{x}{2} - \frac{1}{4}\sin 2x + C.$

5. $\int \cos^2 x dx = \frac{x}{2} + \frac{1}{4}\sin 2x + C.$

(c) *In general, when n is any integer,* let

$$u = \sin^{n-1} x, \quad dv = \sin x dx.$$

Substituting in $\int u dv = uv - \int v du$, we have

$$\int \sin^n x dx = -\sin^{n-1} x \cos x + (n-1) \int \cos^2 x \sin^{n-2} x dx$$

$$= -\sin^{n-1} x \cos x + (n-1) \int (1 - \sin^2 x) \sin^{n-2} x dx$$

$$= -\sin^{n-1} x \cos x + (n-1) \int \sin^{n-2} x dx - (n-1)$$

$$\int \sin^n x dx.$$

Transposing the last term to the first member,

$$\int \sin^n x dx = -\frac{\sin^{n-1} x \cos x}{n} + \frac{n-1}{n} \int \sin^{n-2} x dx.$$

The integration is thus finally made to depend upon $\int dx = x$ if n is even, or upon $\int \sin x dx = -\cos x$ if n is odd.

In like manner,

$$\int \cos^n x dx = \frac{\cos^{n-1} x \sin x}{n} + \frac{n-1}{n} \int \cos^{n-2} x dx,$$

the integration depending on $\int dx = x$ if n is even, or upon $\int \cos x dx = \sin x$ if n is odd.

6. $\int \sin^4 x dx = -\dfrac{\sin^3 x \cos x}{4} + \dfrac{3}{4} \int \sin^2 x dx$

$$= -\frac{\sin^3 x \cos x}{4} + \frac{3}{4} \left[-\frac{\sin x \cos x}{2} + \frac{1}{2} \int dx \right]$$

$$= -\frac{\sin^3 x \cos x}{4} - \frac{3}{8} \sin x \cos x + \frac{3}{8} x + C.$$

7. $\int \cos^4 x\,dx = \dfrac{\cos^3 x \sin x}{4} + \frac{3}{8} \sin x \cos x + \frac{3}{8} x + C.$

8. $\int \cos^6 x\,dx = \dfrac{\cos^5 x \sin x}{6} + \frac{5}{24} \cos^3 x \sin x$
$$+ \tfrac{15}{48} \sin x \cos x + \tfrac{15}{16} x + C.$$

II. $\dfrac{dx}{\sin^n x}$ and $\dfrac{dx}{\cos^n x}.$

(a) When n is an even integer, we may write

$$\frac{dx}{\sin^n x} = \operatorname{cosec}^{n-2} x \operatorname{cosec}^2 x\,dx = (1+\cot^2 x)^{\frac{n-2}{2}} \operatorname{cosec}^2 x\,dx,$$

and $\quad \dfrac{dx}{\cos^n x} = (1 + \tan^2 x)^{\frac{n-2}{2}} \sec^2 x\,dx.$

9. $\int \dfrac{dx}{\sin^6 x} = \int \operatorname{cosec}^4 x \operatorname{cosec}^2 x\,dx = \int (1+\cot^2 x)^2 \operatorname{cosec}^2 x\,dx$
$$= -\cot x - \tfrac{2}{3} \cot^3 x - \tfrac{1}{5} \cot^5 x + C.$$

10. $\int \dfrac{dx}{\cos^4 x} = \tan x + \frac{1}{3} \tan^3 x + C.$

11. $\int \dfrac{dx}{\cos^6 x} = \tan x + \frac{2}{3} \tan^3 x + \frac{1}{5} \tan^5 x + C.$

(b) When n is 1, we have

12. $\int \dfrac{dx}{\sin x} = \int \dfrac{dx}{2 \sin \frac{x}{2} \cos \frac{x}{2}} = \int \dfrac{\dfrac{dx}{2 \cos^2 \frac{x}{2}}}{\dfrac{\sin \frac{x}{2}}{\cos \frac{x}{2}}} = \int \dfrac{\frac{1}{2} \sec^2 \frac{x}{2}\,dx}{\tan \frac{x}{2}}$

$$= \log \tan \frac{x}{2} + C.$$

13. $\int \dfrac{dx}{\cos x} = \int \dfrac{dx}{\sin\left(\frac{\pi}{2} - x\right)} = -\log \tan\left(\frac{\pi}{4} - \frac{x}{2}\right) + C,$ by Ex. 12.

(c) *In general, when n is any integer,*

$$\int \frac{dx}{\sin^n x} = \int \frac{\cos^2 x + \sin^2 x}{\sin^n x} dx = \int \cos x \frac{\cos x dx}{\sin^n x} + \int \frac{dx}{\sin^{n-2} x}.$$

Let $u = \cos x, \quad dv = \dfrac{\cos x dx}{\sin^n x}.$

Substituting in $\int u dv = uv - \int v du$, we have

$$\int \cos x \frac{\cos x dx}{\sin^n x} = -\frac{\cos x}{(n-1)\sin^{n-1} x} - \frac{1}{n-1}\int \frac{dx}{\sin^{n-2} x}.$$

Hence $\displaystyle\int \frac{dx}{\sin^n x} = -\frac{\cos x}{(n-1)\sin^{n-1} x} + \frac{n-2}{n-1}\int \frac{dx}{\sin^{n-2} x}.$

The integration is thus finally made to depend upon

$$\int \frac{dx}{\sin^2 x} = \int \operatorname{cosec}^2 x dx = -\cot x,$$

if n is even, or upon

$$\int \frac{dx}{\sin x} = \log \tan \frac{x}{2} \text{ (Ex. 12)},$$

if n is odd.

In like manner,

$$\int \frac{dx}{\cos^n x} = \frac{\sin x}{(n-1)\cos^{n-1} x} + \frac{n-2}{n-1}\int \frac{dx}{\cos^{n-2} x},$$

the integration depending upon

$$\int \frac{dx}{\cos^2 x} = \int \sec^2 x dx = \tan x,$$

if n is even, or upon

$$\int \frac{dx}{\cos x} = -\log \tan\left(\frac{\pi}{4} - \frac{x}{2}\right) \text{(Ex. 13)},$$

if n is odd.

14. $\int \dfrac{dx}{\sin^5 x} = -\dfrac{\cos x}{4\sin^4 x} + \dfrac{3}{4}\int \dfrac{dx}{\sin^3 x}$

$= -\dfrac{\cos x}{4\sin^4 x} + \dfrac{3}{4}\left(-\dfrac{\cos x}{2\sin^2 x} + \dfrac{1}{2}\int \dfrac{dx}{\sin x}\right)$

$= -\dfrac{\cos x}{4\sin^4 x} - \dfrac{3}{8}\dfrac{\cos x}{\sin^2 x} + \dfrac{3}{8}\log \tan\dfrac{x}{2} + C.$

15. $\int \dfrac{dx}{\cos^3 x} = \dfrac{\sin x}{2\cos^2 x} - \dfrac{1}{2}\log \tan\left(\dfrac{\pi}{4} - \dfrac{x}{2}\right) + C.$

III. $\dfrac{\sin^n x\,dx}{\cos^m x}$ and $\dfrac{\cos^n x\,dx}{\sin^m x}.$

(a) When $n = 1$, we have directly, by Form 1,

16. $\dfrac{\sin x}{\cos^7 x}dx = -\int (\cos x)^{-7}(-\sin x\,dx) = \dfrac{1}{6}\dfrac{1}{\cos^6 x} + C.$

17. $\int \dfrac{\cos x}{\sin^4 x}dx = -\dfrac{1}{3}\dfrac{1}{\sin^3 x} + C.$

(b) When $n - m$ is *negative and even*, Form 1 applies if we write

$$\dfrac{\sin^n x\,dx}{\cos^m x} = \tan^n x \sec^{m-n} x\,dx,$$

and $$\dfrac{\cos^n x\,dx}{\sin^m x} = \cot^n x \operatorname{cosec}^{m-n} x\,dx.$$

18. $\int \dfrac{\sin^3 x}{\cos^7 x}dx = \int \tan^5 x \sec^2 x\,dx = \tfrac{1}{6}\tan^6 x + C.$

19. $\int \dfrac{\cos^3 x}{\sin^9 x}dx = \int \cot^3 x \operatorname{cosec}^6 x\,dx$

$= \int \cot^3 x (1 + \cot^2 x)^2 \operatorname{cosec}^2 x\,dx$

$= -\tfrac{1}{4}\cot^4 x - \tfrac{1}{3}\cot^6 x - \tfrac{1}{8}\cot^8 x + C.$

20. $\int \dfrac{\sin^2 x\,dx}{\cos^4 x} = \tfrac{1}{3}\tan^3 x + C.$

21. $\displaystyle\int \frac{\cos^2 x\, dx}{\sin^4 x} = -\tfrac{1}{3}\cot^3 x + C.$

22. $\displaystyle\int \frac{\sin^4 x}{\cos^{10} x}\, dx = \tfrac{1}{5}\tan^5 x + \tfrac{2}{7}\tan^7 x + \tfrac{1}{9}\tan^9 x + C.$

(c) *When $n - m$ is negative and odd, if n is odd*, we have

$$\frac{\sin^n x}{\cos^m x}\, dx = \tan^n x \sec^{m-n} x\, dx$$

$$= (\sec^2 x - 1)^{\frac{n-1}{2}} \sec^{m-n-1} x \tan x \sec x\, dx,$$

to which Form 1 applies, and $\dfrac{\cos^n x\, dx}{\sin^m x}$ may be treated in a similar manner.

23. $\displaystyle\int \frac{\sin^5 x}{\cos^{10} x}\, dx = \int \tan^5 x \sec^5 x\, dx$

$$= \int (\sec^2 x - 1)^2 \sec^4 x \tan x \sec x\, dx$$

$$= \int (\sec^8 x - 2\sec^6 x + \sec^4 x)\tan x \sec x\, dx$$

$$= \tfrac{1}{9}\sec^9 x - \tfrac{2}{7}\sec^7 x + \tfrac{1}{5}\sec^5 x + C.$$

24. $\displaystyle\int \frac{\sin^3 x\, dx}{\cos^6 x} = \tfrac{1}{5}\sec^5 x - \tfrac{1}{3}\sec^3 x + C.$

25. $\displaystyle\int \frac{\cos^3 x}{\sin^4 x}\, dx = -\tfrac{1}{3}\operatorname{cosec}^3 x + \operatorname{cosec} x + C.$

(d) *When $n - m$ is positive*, resort must be had to integration by parts. *When, however, $m - n = 1$, and n is odd*,

26. $\displaystyle\int \frac{\sin^3 x\, dx}{\cos^2 x} = \int (1 - \cos^2 x)\frac{\sin x}{\cos^2 x}\, dx = \sec x + \cos x + C.$

27. $\displaystyle\int \frac{\cos^3 x\, dx}{\sin^2 x} = -\operatorname{cosec} x - \sin x + C.$

28. $\displaystyle\int \frac{\sin^7 x\, dx}{\cos^6 x} = \frac{1}{5\cos^5 x} - \frac{1}{\cos^3 x} + \frac{3}{\cos x} + \cos x + C.$

IV. **tan^m $x dx$** and **cot^m $x dx$.** These forms can be integrated directly, *when m is integral and positive*, by placing

$$\tan^m x dx = (\sec^2 x - 1)\tan^{m-2} x dx,$$

and
$$\cot^m x dx = (\operatorname{cosec}^2 x - 1)\cot^{m-2} x dx.$$

29. $\displaystyle\int \tan^2 x dx = \int (\sec^2 x - 1)dx = \tan x - x + C.$

30. $\displaystyle\int \tan^3 x dx = \int (\sec^2 x - 1)\tan x dx = \tfrac{1}{2}\tan^2 x - \int \tan x dx$

$$= \tfrac{1}{2}\tan^2 x - \int \frac{\sin x}{\cos x}dx = \tfrac{1}{2}\tan^2 x + \log \cos x + C.$$

31. $\displaystyle\int \tan^4 x dx = \int (\sec^2 x - 1)\tan^2 x dx = \tfrac{1}{3}\tan^3 x - \int \tan^2 x dx$

$$= \tfrac{1}{3}\tan^3 x - \tan x + x + C \text{ (Ex. 29).}$$

32. $\displaystyle\int \tan^5 x dx = \tfrac{1}{4}\tan^4 x - \tfrac{1}{2}\tan^2 x - \log \cos x + C.$

33. $\displaystyle\int \cot^2 x dx = -\cot x - x + C.$

34. $\displaystyle\int \cot^3 x dx = -\tfrac{1}{2}\cot^2 x - \log \sin x + C.$

35. $\displaystyle\int \cot^5 x dx = -\tfrac{1}{4}\cot^4 x + \tfrac{1}{2}\cot^2 x + \log \sin x + C.$

36. $\displaystyle\int (\tan^2 x + \tan^4 x)dx = \int \tan^2 x \sec^2 x dx = \tfrac{1}{3}\tan^3 x + C.$

37. $\displaystyle\int (\tan^6 x + \tan^8 x)dx = \int \tan^6 x \sec^2 x dx = \tfrac{1}{7}\tan^7 x + C.$

And, in like manner, $(\tan^n x + \tan^m x)dx$ when $n - m = 2$.

V. **$x^n \sin (ax)dx$,** and **$x^n \cos (ax)dx$.**

Let $u = x^n$, $dv = \sin (ax)dx$.

Substituting in $\displaystyle\int u dv = uv - \int v du$,

$$\int x^n \sin (ax)dx = -\frac{x^n \cos (ax)}{a} + \frac{n}{a}\int \cos (ax)x^{n-1}dx,$$

the integration finally depending upon

$$\int \cos (ax)dx \text{ or } \int \sin (ax)dx.$$

38. $\displaystyle\int x^3 \cos x dx = x^3 \sin x - 3 \int x^2 \sin x dx$

$$= x^3 \sin x - 3(- x^2 \cos x - 2 \int - x \cos x dx)$$

$$= x^3 \sin x + 3 x^2 \cos x - 6 (x \sin x - \int \sin x dx)$$

$$= x^3 \sin x + 3 x^2 \cos x - 6 x \sin x - 6 \cos x + C.$$

39. $\displaystyle\int x^2 \sin x dx = - x^2 \cos x + 2 x \sin x + 2 \cos x + C.$

40. $\displaystyle\int x \sin (mx)dx = - \frac{x \cos (mx)}{m} + \frac{\sin (mx)}{m^2} + C.$

VI. $e^{ax} \sin^n x dx$, and $e^{ax} \cos^n x dx$.

Let $u = \sin^n x, \; dv = e^{ax}dx.$

Substituting in $\displaystyle\int u dv = uv - \int v du,$

$$\int e^{ax} \sin^n x dx = \frac{\sin^n x e^{ax}}{a} - \frac{n}{a} \int e^{ax} \sin^{n-1} x \cos x dx. \qquad (1)$$

In the last integral let $u = \sin^{n-1} x \cos x, \; dv = e^{ax}dx.$ Then

$$du = (n-1) \sin^{n-2} x \cos^2 x dx - \sin^n x dx$$

$$= (n-1) \sin^{n-2} x (1 - \sin^2 x) dx - \sin^n x dx$$

$$= (n-1) \sin^{n-2} x dx - n \sin^n x dx,$$

$$v = \frac{e^{ax}}{a};$$

and the formula $\displaystyle\int u dv = uv - \int v du$ gives

$$\int e^{ax} \sin^{n-1} x \cos x dx$$

$$= \frac{\sin^{n-1} x \cos x e^{ax}}{a} - \frac{n-1}{a} \int e^{ax} \sin^{n-2} x dx + \frac{n}{a} \int e^{ax} \sin^n x dx.$$

Substituting in (1) and solving for $\int e^{ax} \sin^n x dx$,

$$\int e^{ax} \sin^n x dx$$
$$= \frac{e^{ax} \sin^{n-1} x}{a^2 + n^2} (a \sin x - n \cos x) + \frac{n(n-1)}{a^2 + n^2} \int e^{ax} \sin^{n-2} x dx. \quad (2)$$

By a repetition of this process the integration is made to depend upon the known form $\int e^{ax} dx$, or upon $\int e^{ax} \sin x dx$, which by (2) is $\frac{e^{ax}}{a^2 + 1} (a \sin x - \cos x)$, n being 1. From the form $e^{ax} \cos^n x dx$ we have in like manner.

$$\int e^{ax} \cos^n x dx$$
$$= \frac{e^{ax} \cos^{n-1} x}{a^2 + n^2} (a \cos x + n \sin x) + \frac{n(n-1)}{a^2 + n^2} \int e^{ax} \cos^{n-2} x dx.$$

41. $\int e^{ax} \sin x dx = \frac{e^{ax}}{a^2 + 1} (a \sin x - \cos x) + C.$

42. $\int e^{ax} \cos^2 x dx = \frac{e^{ax} \cos x}{a^2 + 4} (a \cos x + 2 \sin x) + \frac{2 e^{ax}}{a(a^2 + 4)} + C.$

43. $\int e^x \sin^3 x dx = \frac{e^x}{10} (\sin^3 x + 3 \cos^3 x + 3 \sin x - 6 \cos x) + C.$

151. Circular differentials of the forms $f(x) \sin^{-1} x dx$, $f(x) \cos^{-1} x dx$, **etc.,** *in which* $f(x)$ *is an algebraic function.*

Assuming $dv = f(x) dx$, the formula for integration by parts will make the integration depend upon an algebraic form.

EXAMPLES. 1. $\int \sin^{-1} x dx.$

$u = \sin^{-1} x,\ dv = dx,\ du = \dfrac{dx}{\sqrt{1 - x^2}},\ v = x.$ Then

$$\int \sin^{-1} x dx = x \sin^{-1} x - \int \frac{x dx}{\sqrt{1 - x^2}} = x \sin^{-1} x + (1 - x^2)^{\frac{1}{2}} + C.$$

2. $\int \tan^{-1} x dx = x \tan^{-1} x - \frac{1}{2} \log (1 + x^2) + C.$

3. $\int x^2 \cos^{-1}x\,dx = \dfrac{x^3}{3}\cos^{-1}x - \dfrac{\sqrt{1-x^2}}{9}\,(x^2+2) + C.$

4. $\int x \cos^{-1}x\,dx = \frac{1}{2}x^2\cos^{-1}x - \frac{1}{4}x(1-x^2)^{\frac{1}{2}} + \frac{1}{4}\sin^{-1}x + C.$

BY SUBSTITUTION.

This method has been already employed in the rationalization of irrational differentials (Arts. 143–4), and consists in substituting for the variable of the given differential a new variable of which it is a function.

152. Trigonometric functions of the form $\sin^n x \cos^m x\,dx$.

I. Let $\sin x = z$. Then

$$\sin^n x = z^n,\quad \cos^m x = (1-z^2)^{\frac{m}{2}},\quad dx = (1-z^2)^{-\frac{1}{2}}dz.$$

Hence $\int \sin^n x \cos^m x\,dx = \int z^n (1-z^2)^{\frac{m-1}{2}}\,dz,$

or in like manner, writing $\cos x = z$,

$$\int \sin^n x \cos^m x\,dx = \int - z^m (1-z^2)^{\frac{n-1}{2}}\,dz.$$

The given differential may then be integrated whenever the above binomials can be integrated.

EXAMPLES. 1. $\int \sin^4 x\,dx.$ $\sin x = z,\ dx = \dfrac{dz}{\cos x} = \dfrac{dz}{\sqrt{1-z^2}}.$

$\int \sin^4 x\,dx = \int \dfrac{z^4 dz}{\sqrt{1-z^2}} = -\dfrac{z}{4}(z^2+\tfrac{3}{2})\sqrt{1-z^2} + \tfrac{3}{8}\sin^{-1}z + C$

$\hfill$ (Ex. 2, Art. 147)

$\quad = -\dfrac{\cos x}{4}(\sin^3 x + \tfrac{3}{2}\sin x) + \tfrac{3}{8}x + C.$

2. $\int \sin^5 x\,dx = \int \dfrac{z^5 dz}{\sqrt{1-z^2}} = -\left(\dfrac{z^4}{5} + \dfrac{4z^2}{15} + \dfrac{8}{15}\right)\sqrt{1-z^2} + C$

$\hfill$ (Ex. 3, Art. 147.)

$\quad = -\dfrac{\cos x}{5}(\sin^4 x + \tfrac{4}{3}\sin^2 x + \tfrac{8}{3}) + C.$

3. $\int \sin^3 x \cos^2 x\, dx = \int z^2 (1-z^2)^{\frac{1}{2}} dz$ when $\sin x = z.$

$$\int z^2 (1-z^2)^{\frac{1}{2}} dz = -\frac{z(1-z^2)^{\frac{3}{2}}}{4} + \frac{z(1-z^2)^{\frac{1}{2}}}{8} + \frac{1}{8}\sin^{-1}z + C.$$

(Ex. 8, Art. 147.)

Hence $\int \sin^2 x \cos^2 x\, dx = -\dfrac{\sin x \cos^3 x}{4} + \dfrac{\sin x \cos x}{8} + \frac{1}{8}x + C.$

II. *When either m or n is odd*, we may integrate directly by treating the factor whose exponent is odd as in Art. 150, I., (a).

4. $\int \sin^3 x \cos^2 x\, dx = \int (1 - \cos^2 x)\cos^2 x \sin x\, dx$

$$= -\tfrac{1}{3}\cos^3 x + \tfrac{1}{5}\cos^5 x + C.$$

5. $\int \cos^5 x \sin^2 x\, dx = \tfrac{1}{3}\sin^3 x - \tfrac{2}{5}\sin^5 x + \tfrac{1}{7}\sin^7 x + C.$

6. $\int \cos^4 x \sin^3 x\, dx = -\tfrac{1}{5}\cos^5 x + \tfrac{1}{7}\cos^7 x + C.$

7. $\int \sin x \cos^2 x\, dx = -\tfrac{1}{3}\cos^3 x + C.$

8. $\int \cos x \sin^5 x\, dx = \tfrac{1}{6}\sin^6 x + C,$

form 1 applying when n or m is 1.

9. $\int \dfrac{dx}{\sin x \cos x} = \int \dfrac{2\, dx}{\sin 2x} = \log \tan x + C.$ (Ex. 12, Art. 150.)

10. $\int \dfrac{dx}{\sin^2 x \cos^2 x} = \int \dfrac{\sin^2 x + \cos^2 x}{\sin^2 x \cos^2 x} dx = \tan x - \cot x + C.$

11. $\int \sin^3 x \cos^3 x\, dx = \tfrac{1}{4}\sin^4 x - \tfrac{1}{6}\sin^6 x + C.$

153. Many differentials may be integrated by substitution, but no general rule can be given, and the method is best exhibited by examples, of which a few are added.

1. $\int \dfrac{dx}{x(a^3 + x^3)} = \dfrac{1}{3}\int \dfrac{dz}{z(a^3 + z)}$ when $x^3 = z$.

By Art. 139, Case 1,

$$\int \frac{dz}{z(a^3 + z)} = \frac{1}{a^3}\log \frac{z}{a^3 + z}.$$

Hence $\qquad \int \dfrac{dx}{x(a^3 + x^3)} = \dfrac{1}{3a^3}\log \dfrac{x^3}{a^3 + x^3} + C.$

2. $\int \dfrac{dx}{x\sqrt{1 + x + x^2}} = -\log \dfrac{2 + x + 2\sqrt{x^2 + x + 1}}{x}.$

Put $x = \dfrac{1}{y}$; the differential may then be integrated by Art. 144, I.

3. $\int \dfrac{1 + x^2}{(1 + x)^2} e^x dx = e^x\left(\dfrac{x - 1}{x + 1}\right) + C.$

Put $\qquad 1 + x = z.$

Then $\qquad \int \dfrac{1 + x^2}{(1 + x)^2} e^x dx = \dfrac{1}{e}\left(e^z + 2\int \dfrac{e^z dz}{z^2} - 2\int \dfrac{e^z dz}{z}\right).$ (1)

Placing $u = e^z$ and $dv = z^{-2}dz$, and applying the formula for integration by parts to $\int \dfrac{e^z dz}{z^2}$, we have

$$\int \frac{e^z dz}{z^2} = -\frac{e^z}{z} + \int \frac{e^z dz}{z}.$$

Substituting this value in (1), we have the above result.

4. $\int \dfrac{dx}{x}\sqrt{1 + \log x} = \dfrac{2}{3}(1 + \log x)^{\frac{3}{2}} + C.$ Let $1 + \log x = z.$

5. $\int \dfrac{x^2 dx}{\sqrt{1 - x^2}} = -\dfrac{1}{2}\cos^{-1}x - \dfrac{x\sqrt{1 - x^2}}{2} + C.$ Let $x = \cos z.$

6. $\int \dfrac{dx}{x^2(a + bx)^2} = \dfrac{2b}{a^3}\log \dfrac{a + bx}{x} - \dfrac{a + 2bx}{a^2 x(a + bx)} + C.$

Put $x = \dfrac{1}{z}$, whence $\dfrac{dx}{x^2(a + bx)^2} = -\dfrac{z^2 dz}{(az + b)^2}.$

In the latter let $az + b = y$, and it becomes $-\dfrac{1}{a^3}\dfrac{(y-b)^2}{y^2}dy$.

7. $x^3(a-x^2)^{\frac{1}{2}}dx$. Put $x^2 = a - z^2$.

BY SERIES.

154. When the given differential can be expanded into a converging series, its integral may be found by integrating each term of the series. The integral thus obtained will be in the form of a series, and therefore integration by series affords a method of developing a function where the development of the derivative is known.

EXAMPLES.

1. $\displaystyle\int\sqrt{x-x^5}\,dx = \int\sqrt{x}(1-x^2)^{\frac{1}{2}}dx$

$$= \int\sqrt{x}(1-\frac{x^2}{2}-\frac{x^4}{8}-\frac{x^6}{16}\cdots)dx$$

$$= \tfrac{2}{3}x^{\frac{3}{2}} - \tfrac{1}{7}x^{\frac{7}{2}} - \tfrac{1}{44}x^{\frac{11}{2}} - \tfrac{1}{120}x^{\frac{15}{2}}\cdots + C.$$

2. $\displaystyle\int\frac{e^x}{\cos x}dx = \int(1+x+x^2+\tfrac{2}{3}x^3+\tfrac{1}{2}x^4\cdots)dx$

$$= x + \frac{x^2}{2} + \frac{x^3}{3} + \frac{x^4}{6} + \frac{x^5}{10}\cdots + C.$$

See Ex. 18, Art. 72.

3. Develop $\log(1+x)$.

$$\log(1+x) = \int\frac{dx}{1+x} = \int(1+x)^{-1}dx$$

$$= \int(1-x+x^2-x^3\cdots)dx$$

$$= x - \frac{x^2}{2} + \frac{x^3}{3} - \frac{x^4}{4}\cdots + C.$$

4. Develop $\sin^{-1}x$.

$$\sin^{-1}x = \int\frac{dx}{\sqrt{1-x^2}} = \int(1+\tfrac{1}{2}x^2+\tfrac{3}{8}x^4+\tfrac{15}{48}x^6+\cdots)dx$$

$$= x + \tfrac{1}{6}x^3 + \tfrac{3}{40}x^5 + \tfrac{5}{112}x^7\cdots + C.$$

REMARK. The process of integration is the inverse of that of differentiation; but it does not follow that, because we can differentiate every integral, we can integrate every differential. Suppose, for example, the given function be x^n; its differential is $nx^{n-1}dx$. Now, in order that the differential of x^n should assume the form $\dfrac{1}{x}$, we must have $n-1=-1$, or $n=0$; in which case $x^n = 1$, which has no differential. That is, the algebraic function x^n cannot give rise to a differential of the form $\dfrac{dx}{x}$; nor can any other known function except $\log x$. It is evident, therefore, that, before the invention of logarithms and the investigation of their properties, the operation indicated by $\displaystyle\int\dfrac{dx}{x}$ would have been impossible. The transcendental functions $\sin^{-1}x$, $\tan^{-1}x$, etc., whose differentials $\dfrac{dx}{\sqrt{1-x^2}}$, $\dfrac{dx}{1+x^2}$, etc., are algebraic functions, are further illustrations of the fact that the integration of algebraic differentials may involve transcendental, or higher, functions. The integration, therefore, of such forms as do not arise by the differentiation of the known functions cannot be effected until new functions corresponding to these forms have been invented.

MISCELLANEOUS EXAMPLES.

Integrate :

1. $\dfrac{1-x^n}{1-x}\,dx.$

2. $\dfrac{xdx}{\sqrt{a^4-x^4}}.$

3. $\dfrac{e^x xdx}{(1+x)^2}.$

4. $\dfrac{\sqrt{x}+1}{\sqrt{x}-1}\,dx.$

5. $\dfrac{1+2x\cos^2 x}{\cos x \sin x + x^2\cos^2 x}\,dx.$

6. $\dfrac{xdx}{(1-x)^3}.$

7. $x\tan^{-1}xdx.$

8. $\dfrac{bdx}{\sqrt{c^2-a^2-2abx-b^2x^2}}.$

9. $\dfrac{3\,dx}{\sqrt{3-6x-9x^2}}$.

10. $\dfrac{x^2 dx}{1+x^4}$.

11. $\dfrac{dx}{\sin x \cos^2 x}$.

12. $\dfrac{x^2-1}{x^4+x^2+1}\,dx$.

13. $\dfrac{dx}{a^4-x^4}$.

14. $\dfrac{x\,dx}{x^4-x^2-2}$.

15. $\dfrac{x\,dx}{x^4-a^4}$.

16. $\dfrac{5x-2}{x^3+6x^2+8x}\,dx$.

17. $\dfrac{dx}{a^2\cos^2 x + b^2\sin^2 x}$.

18. $\dfrac{x^{n-1}dx}{(a+bx^n)^m}$.

19. $\dfrac{m-x}{\sqrt{2mx-x^2}}\,dx$.

20. $\dfrac{a+bx}{c^2+x^2}\,dx$.

21. $\dfrac{m\,dx}{a+bx^2}$.

22. $\dfrac{mx\,dx}{a+bx^2}$.

23. $\dfrac{b+2cx}{(a+bx+cx^2)^n}\,dx$.

24. $\cos 2x\,dx$.

25. $\dfrac{\sin nx\,dx}{m-\cos nx}$.

26. $\dfrac{\sec^2 x\,dx}{m-n\tan x}$.

27. $\dfrac{nx^{n-1}dx}{\sqrt{a^{2n}-x^{2n}}}$.

28. $\dfrac{\cos x\,dx}{a^2+\sin^2 x}$.

29. $e^{x-n}\,dx$.

30. $\dfrac{(x-a)dx}{(x-a)^2+(x+a)^2}$.

31. $\dfrac{(x-a)dx}{(x-a)^2 \pm (x+a)^2}$.

32. $\dfrac{dx}{x(x+1)^2}$.

33. $\dfrac{dx}{x^2(x-1)^2}$.

34. $\dfrac{1+x^{-\frac{1}{2}}}{1-x^{-\frac{1}{2}}}\,dx$.

35. $x^3\sqrt{1-x^2}\,dx$.

36. $x^6\sqrt{1+x^2}\,dx$.

37. $e^{2x}\sin 3x\,dx$.

38. $\dfrac{dx}{x^4\sqrt{1+x^2}}$.

SUCCESSIVE INTEGRATION.

155. Successive differentials obtained on the hypothesis that the variable is equicrescent are readily integrated by the preceding methods, the differential of the variable being constant.

EXAMPLES.   1. Given $d^2y = 10\,x^2dx^2$, to find y.

$$\frac{d^2y}{dx} = 10\,x^2dx; \text{ integrating, } \frac{dy}{dx} = \tfrac{10}{3}\,x^3 + C'.$$

$$dy = \tfrac{10}{3}\,x^3dx + C'dx; \text{ integrating, } y = \tfrac{5}{6}x^4 + C'x + C''.$$

2. Given $\dfrac{d^3y}{dx^3} = \cos x$, to find y.

$$\frac{d^3y}{dx^2} = \cos x\,dx; \therefore \frac{d^2y}{dx^2} = \sin x + C'.$$

$$\frac{d^2y}{dx} = \sin x\,dx + C'dx; \therefore \frac{dy}{dx} = -\cos x + C'x + C''.$$

$$dy = -\cos x\,dx + C'x\,dx + C''dx;$$

$$\therefore y = -\sin x + \frac{C'x^2}{2} + C''x + C'''.$$

3. Given $\dfrac{d^2y}{dx^2} = 0$, to find y.

$$\frac{d^2y}{dx} = 0; \therefore \frac{dy}{dx} = C'. \quad dy = C'dx; \therefore y = C'x + C''.$$

4. Given $d^4y = \sin x\,dx^4$, to find y.

5. Given $d^2s = -g\,dt^2$, to find s.

6. Given $\dfrac{d^3y}{dx^3} = -\dfrac{1}{x^3}$, to find y.

THE CONSTANT OF INTEGRATION.

156. All the integrals thus far obtained contain the indeterminate constant C, and are called **indefinite integrals.**

Integrals from which the constant has been eliminated, or for which its value has been determined, are called **definite integrals.**

157. Definite integrals. The two methods of disposing of the constant of integration C are best explained by an illustration of the processes. Let it be required to find the plane area $OM'N'$ between the parabola OM', the ordinate $M'N'$, and the axis of X. This area may be regarded as generated by the motion of the ordinate PD from left to right. If this area be represented by z, dz will represent what its change would be in any interval of time, dt, if its rate of increase remained uniformly the same during that interval. But if the rate of z becomes constant at any instant, that is, at any value PD of y, its increase for any interval dt will be represented by $PQRD = PD \times DR = ydx$; $DR = dx$ being the corresponding differential of x. Hence $dz = ydx$, and

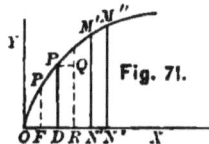

$$z = \int ydx. \tag{1}$$

Substituting the value $dx = \dfrac{y}{p} dy$ from the equation of the parabola $y^2 = 2px$,

$$dz = \frac{y^2}{p} dy \tag{2}$$

and

$$z = \frac{1}{p} \int y^2 dy = \frac{y^3}{3p} + C. \tag{3}$$

FIRST METHOD. Evidently the area generated cannot be definitely expressed until we assume some initial position of PD

as an origin from which to estimate it. If we reckon the area from the ordinate through the focus F, then $z = 0$ when $y = FP' = p$, and (3) gives $C = -\dfrac{p^2}{3}$, and the definite integral is

$$z = \frac{y^3}{3p} - \frac{p^2}{3},$$

which gives the area, estimated from FP', to any position of y as $M'N'$ when $y' = M'N'$ is substituted for y.

If we reckon the area from O, then $z = 0$ when $y = 0$, and (3) gives $C = 0$, the definite integral being

$$z = \frac{y^3}{3p},$$

which gives the area, estimated from O, to any position $M'N'$ of y, when $y' = M'N'$ is substituted for y.

Hence the value of C may be found whenever we know the value of the function for a particular value of the variable ; and it is evident that this will be the case in all problems like the above, in which the origin from which the magnitude is to be estimated may be arbitrarily chosen.

SECOND METHOD. If we substitute any value of y, as $y'' = M''N''$, in (3),

$$z'' = \frac{y''^3}{3p} + C$$

is the area generated while the ordinate is moving to the position $M''N''$. Substituting $y' = M'N'$,

$$z' = \frac{y'^3}{3p} \stackrel{.}{\cdot} C$$

is the area generated while the ordinate is moving to the position $M'N'$. Hence

$$z'' - z' = \left(\frac{y''^3}{3p} + C\right) - \left(\frac{y'^3}{3p} + C\right) = \frac{y''^3 - y'^3}{3p}$$

is the area generated in moving from $M'N'$ to $M''N''$, and is independent of any initial position of the ordinate. In other words, the area is increasing at the rate $\dfrac{dz}{dt} = \dfrac{y^2}{p}\dfrac{dy}{dt}$, and the area generated at that rate while y passes from the value y' to the value y'' is found by substituting these values in (3) and taking the difference of the results. In this way C is eliminated, the process being called **integration between limits**.

The symbol for the integral between the limits y' and y'' is $\displaystyle\int_{y'}^{y''} \phi(y)\,dy$, y'' being the *superior* and y' the *inferior* limit; and it indicates that in the integral of $\phi(y)\,dy$, y'' and y' are to be substituted for the variable in succession, and the latter result subtracted from the former. It is to be observed that the two methods are essentially the same, for in the first the inferior limit is assumed in determining the value of C, and the superior limit is the value subsequently assigned to the variable in the definite integral.

The constants introduced in successive integration are readily determined from the conditions of the problem if the latter is a determinate one.

Thus, suppose a body starts *from rest* with a constant acceleration m in a right line. Taking the axis of X coincident with the rectilinear path, we have (Art. 59),

$$\frac{d^2x}{dt^2} = m.$$

Multiplying by dt and integrating,

$$\frac{dx}{dt} = v = mt + C. \tag{1}$$

Reckoning t from the instant the body starts, we have, by condition, $v = 0$ when $t = 0$; hence $C = 0$, and

$$\frac{dx}{dt} = v = mt. \tag{2}$$

Integrating again,

$$x = \frac{mt^2}{2} + C'. \tag{3}$$

Reckoning x from the initial position of the body, $x = 0$ when $t = 0$; hence $C' = 0$, and

$$x = \frac{mt^2}{2}. \tag{4}$$

Eliminating t between (2) and (4), we have for the equations of motion,

$$v = mt, \quad x = \frac{mt^2}{2}, \quad v = \sqrt{2\,mx},$$

from which we may find the position of the body at any time, and its velocity at any time or at any point of the path.

Had the body an initial velocity v_0 when $x = t = 0$, we should have had from (1), $C = v_0$, and therefore

$$\frac{dx}{dt} = v = mt + v_0;$$

whence, integrating again,

$$x = \frac{mt^2}{2} + v_0 t + C',$$

in which $C' = 0$, since $x = 0$ when $t = 0$. The equations of motion in this case would be

$$v = mt + v_0, \quad x = \frac{mt^2}{2} + v_0 t, \quad v^2 = v_0^2 + 2\,mx.$$

And, in general, the equations of motion can be found whenever the position and velocity of the body at any instant is known.

CHAPTER VIII.

GEOMETRICAL APPLICATIONS.

158. Determination of the equations of curves.

1. *To find the equation of the curve whose normal is constant.*

Let R = length of normal. Then (Art. 27, Ex. 21),

$$R = y\sqrt{1 + \left(\frac{dy}{dx}\right)^2},$$

or $\quad x = \pm \int y(R^2 - y^2)^{-\frac{1}{2}}\,dy = \mp (R^2 - y^2)^{\frac{1}{2}} + C.$ $\quad\quad$ (1)

In this, as in all like cases, the fact that the position of the origin of coordinates is arbitrary enables us to determine C. Thus if we assume that the origin is so chosen that $y = R$ when $x = 0$, then, from (1), $C = 0$. Hence $x = \mp \sqrt{R^2 - y^2}$, or, squaring, $x^2 + y^2 = R^2$; the curve being a circle, and the constant of integration being determined upon the condition that the origin is at the centre.

2. *To find the curve whose subtangent is constant.*

$y\dfrac{dx}{dy} = m$; hence $x = \log_a y + C$, or $x = \log_a y$ if $x = 0$ when $y = 1$.

See Ex. 8, Art. 30.

3. *To find the curve whose subnormal is constant.*

$y\dfrac{dy}{dx} = p.$ Hence $y^2 = 2px$ if $x = 0$ when $y = 0$.

4. *To find the curve whose subnormal is always equal to the abscissa of the point of contact.*

An equilateral hyperbola.

5. *To find the curve whose tangent is constant.*

$$y\sqrt{1+\left(\frac{dx}{dy}\right)^2} = a; \quad \text{whence } dx = \mp \frac{(a^2 - y^2)^{\frac{1}{2}}}{y} dy.$$

Taking the negative sign, that is, the case in which y is a decreasing function of x,

$$x = -\int \frac{(a^2 - y^2)^{\frac{1}{2}}}{y} dy = -\int \left(\frac{a^2}{y(a^2 - y^2)^{\frac{1}{2}}} - \frac{y}{(a^2 - y^2)^{\frac{1}{2}}} \right) dy$$

$$= a \log \frac{a + (a^2 - y^2)^{\frac{1}{2}}}{y} - (a^2 - y^2)^{\frac{1}{2}} + C. \qquad \text{(Ex. 5, Art. 145.)}$$

Assuming the origin so that $x = 0$ when $y = a$, we have $C = 0$. The curve is called the **tractrix**, and is shown in the figure.

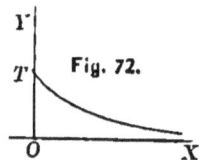
Fig. 72.

6. *Find the curve whose polar subtangent is constant.*

$$r^2 \frac{d\theta}{dr} = a \quad \text{(Art. 120).} \quad \text{The reciprocal spiral.}$$

7. *Find the curve whose polar subnormal is constant.*

159. Rectification of plane curves. The process of finding the length of a curve is called **rectification**.

I. *To rectify* $f(x, y) = 0$. From Art. 25, $ds = \sqrt{dx^2 + dy^2}$; hence

$$s = \int \sqrt{dx^2 + dy^2}. \qquad (1)$$

II. *To rectify* $f(r, \theta) = 0$. From Art. 120, $ds = \sqrt{dr^2 + r^2 d\theta^2}$; hence

$$s = \int \sqrt{dr^2 + r^2 d\theta^2}. \qquad (2)$$

By substituting the value of dx, or of dy, from the equation of the curve in (1), s may be expressed in terms of a single variable and its value found when the integration is possible. If the curve is given by its polar equation, the second form of s is in like manner expressed in terms of a single variable.

EXAMPLES. Rectify the following curves :

1. The semi-cubical parabola $y^2 = ax^3$.

$$dy = \tfrac{3}{2}\sqrt{ax}\,dx;$$

hence $\qquad s = \tfrac{1}{2}\int (4 + 9ax)^{\frac{1}{2}}dx = \dfrac{1}{27\,a}(4 + 9ax)^{\frac{3}{2}} + C.$

Estimating the length from the vertex, $s = 0$ when $x = 0$;

$$\therefore\; C = -\frac{8}{27\,a}, \text{ and } s = \frac{1}{27\,a}[(4 + 9ax)^{\frac{3}{2}} - 8],$$

which is the length of the curve from the vertex to any point whose abscissa is x.

2. The cycloid $x = r\,\mathrm{vers}^{-1}\dfrac{y}{r} - \sqrt{2ry - y^2}$.

$$dx = \sqrt{\frac{y}{2ry}}\,dy;$$

hence $\qquad s = \sqrt{2r}\int (2r - y)^{-\frac{1}{2}}dy = -2\sqrt{2r}(2r - y)^{\frac{1}{2}} + C.$

Estimating from the origin, $s = 0$ when $y = 0$; whence

$$C = 4r,$$

and $\qquad s = -2\sqrt{2r}(2r - y)^{\frac{1}{2}} + 4r\big]_{y=2r} = 4r.$

Hence the entire length of one branch is $8r$.

3. The parabola $y^2 = 2px$.

$$s = \frac{1}{p}\int (p^2 + y^2)^{\frac{1}{2}}dy = \frac{y}{2p}\sqrt{p^2 + y^2} + \frac{p}{2}\log(y + \sqrt{y^2 + p^2}) + C.$$

(See Art. 147, II., the illustrative example.)

Estimating from the vertex,

$$C = -\frac{p}{2}\log p,$$

and

$$s = \frac{y}{2p}\sqrt{y^2 + p^2} + \frac{p}{2}\log\frac{y + \sqrt{y^2 + p^2}}{p}.$$

4. The catenary $y = \frac{c}{2}(e^{\frac{z}{c}} + e^{-\frac{z}{c}})$.

Estimating the arc from the point for which $x = 0$,

$$s = \frac{c}{2}(e^{\frac{z}{c}} - e^{\frac{z}{c}}).$$

5. The hypocycloid $x^{\frac{2}{3}} + y^{\frac{2}{3}} = a^{\frac{2}{3}}$.   *Ans.* $6a$.

6. Determine the length of the tractrix.
From Ex. 5, Art. 158,

$$dx^2 = \frac{a^2 - y^2}{y^2}dy^2.$$

Hence $s = \int \sqrt{dx^2 + dy^2} = -\int \frac{a}{y}dy = -a\log y + C,$

taking the negative sign as s is a decreasing function of y.
(Fig. 72.)   Estimating the arc from T, $s = 0$ when $y = a$;
hence $C = a\log a$, and $s = a\log\frac{a}{y}$.

7. Determine the length of the ellipse.
Using the central form of the equation in terms of the
eccentricity,

$$y^2 = (1 - e^2)(a^2 - x^2), \qquad dy^2 = \frac{(1 - e^2)x^2 dx^2}{a^2 - x^2};$$

hence

$$s = \int \sqrt{dx^2 + dy^2} = \int \sqrt{\frac{a^2 - e^2x^2}{a^2 - x^2}}dx = \int \frac{dx}{\sqrt{a^2 - x^2}}(a^2 - e^2x^2)^{\frac{1}{2}};$$

and for the length of the entire curve,

$$s = 4 \int_0^a \frac{dx}{\sqrt{a^2 - x^2}} \left(a - \frac{e^2 x^2}{2a} - \frac{e^4 x^4}{2 \cdot 4 a^3} - \frac{3 e^6 x^6}{2 \cdot 4 \cdot 6 a^5} - \cdots \right)$$

(Ex. 25, Art. 72.)

$$= 4a \int_0^a \frac{dx}{\sqrt{a^2 - x^2}} - 2 \frac{e^2}{a} \int_0^a \frac{x^2 dx}{\sqrt{a^2 - x^2}} - \frac{e^4}{2 a^3} \int_0^a \frac{x^4 dx}{\sqrt{a^2 - x^2}}$$

$$- \frac{e^6}{4} \int_0^a \frac{x^6 dx}{\sqrt{a^2 - x^2}} \cdots$$

$$= 2 \pi a \left(1 - \frac{e^2}{2^2} - \frac{3 e^4}{2^2 4^2} - \frac{3^2 \cdot 5 e^6}{2^2 \cdot 4^2 \cdot 6^2} - \cdots \right).$$

The second and third of the above integrals are given in Art. 147, Exs. 1 and 2.

8. The logarithmic spiral $r = a^\theta$, a being the base and m the modulus of the logarithmic system.

$$dr = \frac{a^\theta}{m} d\theta; \quad s = \int \left(\frac{a^{2\theta}}{m^2} + a^{2\theta} \right)^{\frac{1}{2}} d\theta = (1 + m^2)^{\frac{1}{2}} r + C \Big]_0^1 = \sqrt{1 + m^2},$$

the length from the point for which $r = 1$ to the pole.
The corresponding arc of the Naperian spiral $= \sqrt{2}$.

9. The spiral of Archimedes, $r = a\theta$.

$$s = a \int \sqrt{1 + \theta^2} d\theta = \frac{1}{a} \int (a^2 + r^2)^{\frac{1}{2}} dr$$

$$= \frac{r(a^2 + r^2)^{\frac{1}{2}}}{2a} + \frac{a}{2} \log \frac{r + \sqrt{a^2 + r^2}}{a}, \quad \text{(Art. 147, II.)}$$

when the arc is estimated from the pole. This is also the length of the arc of the parabola $y^2 = 2ax$ from the vertex to $y = r$ (Ex. 3); hence this spiral is often called the parabolic spiral.

10. $r = a(1 + \cos \theta)$.

$$S = \int (dr^2 + r^2 d\theta^2)^{\frac{1}{2}} = \int \sqrt{2 a^2 (1 + \cos \theta)} d\theta$$

$$= \int \sqrt{4 a^2 \cos^2 \frac{\theta}{2}} d\theta = 2a \int \cos \frac{\theta}{2} d\theta = 4a \sin \frac{\theta}{2} + C.$$

Estimating the area from the point for which $\theta = 0$, we have $C = 0$, and $s = 4\,a\sin\dfrac{\theta}{2}$.

The curve is a cardioide, the polar axis being the axis of symmetry, and its entire length is $8\,a$.

160. Quadrature of plane areas. I. The plane area included between $y = f(x)$ and the axis of X is given (Art. 157) by

$$z = \int y\,dx. \tag{1}$$

In like manner $z' = \int x\,dy$ gives the area between the curve and Y.

If the curve crosses X, y, and therefore $\dfrac{dz}{dx}$, becomes negative, z being a decreasing function of x; hence areas below X must be considered as negative.

II. By the area of a polar curve is meant the area swept over by its radius vector. Thus OPQ is the area of MN between the limits P and Q.

Representing the area by z, its change would evidently become uniform at any value of $r = OP$ if at this value the generating point moved uniformly in the circular arc PP'. Hence if $d\theta = pp'$,

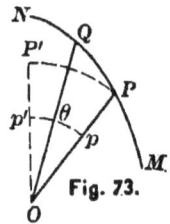

Fig. 73.

$$dz = \text{area } OPP' = \tfrac{1}{2}\,OP \times PP' = \tfrac{1}{2}\,r \cdot r\,d\theta,$$

or

$$z = \tfrac{1}{2}\int r^2 d\theta. \tag{2}$$

The process of finding the area is called **Quadrature.**

EXAMPLES.   1. Determine the area of the parabola $y^2 = 2px$. $dx = \dfrac{y}{p}\,dy$; hence

$$z = \int y\,dx = \frac{1}{p}\int y^2 dy = \frac{y^3}{3p} + C.$$

Estimating the area from the vertex, $z = 0$ when $y = 0$; hence $C = 0$, and $z = \dfrac{y^3}{3p} = \frac{2}{3}xy$, or two-thirds the circumscribing rectangle.

2. Determine the area between $y = \sin x$ and X.

$$z = \int_0^\pi \sin x\, dx = -\cos x \Big]_0^\pi = 2.$$

3. Show that the area between the witch $x^2y = 4\,r^2(2\,r - y)$ and its asymptote is $4\pi r^2$.

$$\int y\,dx = \int \frac{8\,r^3 dx}{x^2 + 4\,r^2} = 4\,r^2 \tan^{-1}\frac{x}{2\,r} \Big]_{-x}^{+x} = 2\,\pi r^2 - (-2\,\pi r^2) = 4\,\pi r^2.$$

4. Show that the area between X and the hyperbola $xy = 1$ from $x = 1$ to $x = x'$ is $\log x'$.

5. Find the area of one branch of the cycloid.

$$\int y\,dx = \int \frac{y^2 dy}{\sqrt{2\,ry - y^2}} = \int y^{\frac{3}{2}}(2\,r - y)^{-\frac{1}{2}}dy$$

$$= -\frac{y + 3\,r}{2}(2\,ry - y^2)^{\frac{1}{2}} + \frac{3\,r^2}{2}\text{vers}^{-1}\frac{y}{r} \Big]_0^{2r} = \frac{3}{2}\pi r^2;$$

hence the whole area is $3\pi r^2$. See Ex. 9, Art. 147.

6. Find the area of the circle $x^2 + y^2 = r^2$.

$$\int y\,dx = \int (r^2 - x^2)^{\frac{1}{2}}dx = \frac{x(r^2 - x^2)^{\frac{1}{2}}}{2} + \frac{r^2}{2}\sin^{-1}\frac{x}{r} \Big]_0^r = \frac{\pi r^2}{4};$$

hence the whole area is πr^2. See Ex. 14. Art. 147.

7. Prove that the area of the ellipse $a^2y^2 + b^2x^2 = a^2b^2$ is πab.

8. Show that the area between the cycloid

$$x = 2\,\text{vers}^{-1}\frac{y}{2} - \sqrt{4\,y - y^2}$$

and the parabola $y^2 = \dfrac{8}{\pi}x$ is $\frac{2}{3}\pi$.

The curves intersect at the origin and $x = 2\pi$.

9. Find the area of $y^2 = x^4 + x^5$ on the left of Y (see Fig. 57).

$$\int y\,dx = \int x^2(1+x)^{\frac{1}{2}}dx = 2\int z^2(z^2-1)^2dz$$

$$= \frac{2(1+x)^{\frac{7}{2}}}{7} - \frac{4}{5}(1+x)^{\frac{5}{2}} + \frac{2}{3}(1+x)^{\frac{3}{2}}]^0_{-1} = \frac{16}{105}.$$

See Ex. 7, Art. 143.

10. Show that the area of the loop $a^2y^4 = a^2x^4 - x^6$ is $\frac{4}{5}a^2$.

11. Show that the area of $y(x^2 + a^2) = c^2(a-x)$ from $x = 0$ to $x = a$ is $c^2\left(\frac{1}{2}\log 2 - \frac{\pi}{4}\right).$

12. Prove that the area between the cissoid $y^2 = \dfrac{x^3}{2a-x}$ and its asymptote is $3\pi a^2$. See Ex. 9, Art. 147.

13. Prove that the area of both loops of $y^2 = x^2(1-x^2)^3$ is $\frac{4}{5}$. See Fig. 59.

14. Prove that the area between X and $y = 4x - x^3$ from $x = -2$ to $x = +2$ is 8.

15. The spiral of Archimedes, $r = a\theta$.

$$z = \frac{1}{2}\int r^2d\theta = \frac{a^2}{2}\int \theta^2d\theta = \frac{a^2}{6}\theta^3 = \frac{1}{6}r^2\theta,$$

C being zero if the area is estimated from $\theta = 0$. For $\theta = 2\pi$, $z = \frac{4}{3}\pi r^2$, or the area of the first spire is $\frac{1}{3}$ that of the measuring circle. When $\theta = 4\pi$, $z = \frac{8}{3}\pi r^2$, or the area of the second spire is $\frac{8}{3}\pi r^2 - \frac{2}{3}\pi r^2 = 2\pi r^2$, the first spire having been traced twice.

16. Prove that the area of $r = e^\theta$ is one-fourth the square described on the radius vector.

17. Find the area of the lemniscate $r^2 = a^2\cos 2\theta$.   *Ans.* a^2.

18. Prove that the area of the cardioide $r = a(1+\cos\theta)$ is $\frac{3}{2}\pi a^2$.

19. Prove that the area of the three loops of $r = a \sin 3\theta$ (Fig. 67) is $\frac{1}{4}\pi a^2$.

20. Find the area of the four loops of $r = a \sin 2\theta$ (Fig. 66).

161. Volumes and surfaces of revolution.

Let the curve ON, whose equation is $y = f(x)$, revolve about X as an axis of revolution. The plane area OQR will generate a solid revolution whose surface will be generated by OQ. A plane section PP' perpendicular to X will cut from this solid a circle whose centre is D and radius is $PD = y$. The volume of the solid may be regarded as generated by this variable circle moving with its centre on X. The rate of every point of this generating area is $\dfrac{dx}{dt}$; hence the rate of increase of the

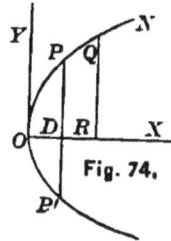

Fig. 74.

volume V is $\dfrac{dV}{dt} = \pi y^2 \dfrac{dx}{dt}$, or

$$V = \int \pi y^2 dx = \pi \int y^2 dx. \qquad (1)$$

The surface S of the solid may be regarded as generated by the circumference of the circle. The rate of every point of this generating circumference is $\dfrac{ds}{dt}$; hence the rate of increase of the surface is $\dfrac{dS}{dt} = 2\pi y \dfrac{ds}{dt}$, or

$$S = \int 2\pi y \, ds = 2\pi \int y \sqrt{dx^2 + dy^2}. \qquad (2)$$

EXAMPLES. 1. Find the volume of the paraboloid of revolution.

$V = \pi \int y^2 dx = \pi \int 2px\,dx = \pi px^2 + C.$ Estimating the volume from the vertex, $V = 0$ when $x = 0$; hence $C = 0$, and

$$V = \pi px^2 = \pi px \frac{y^2}{2p} = \frac{1}{2}\pi y^2 x,$$

or one half the volume of the circumscribing cylinder.

2. Find the volume of the prolate speroid.

$V = \pi \int_0^a \dfrac{b^2}{a^2} (a^2 - x^2) dx = \tfrac{2}{3} \pi b^2 a.$ Hence the whole volume $= \tfrac{2}{3} \pi b^2 (2\,a)$, or two-thirds the circumscribing cylinder.   If $a = b = R$, $V = \tfrac{4}{3} \pi R^3$.

3. Find the volume of the oblate spheroid.

Here $V = \pi \int x^2 dy = \tfrac{2}{3} \pi a^2 (2\,b).$

4. Find the volume generated by the revolution of $y = -\dfrac{b}{a} x + b$ about X.

$$\pi \int_0^a y^2 dx = \tfrac{1}{3} \pi b^2 a.$$

5. Find the volume generated by the revolution of the cycloid about X.

$$\int \pi y^2 dx = \int \pi y^2 \frac{y dy}{\sqrt{2 r y - y^2}} = \pi \int y^3 (2\,ry - y^2)^{-\frac{1}{2}} dy$$

$$= -\pi \frac{2\,y^2 + 5\,r(y + 3\,r)}{6} \sqrt{2\,ry - y^2}$$

$$+ \tfrac{5}{2} \pi r^3 \operatorname{vers}^{-1} \frac{y}{r} + C \Big]_0^{2r} = \tfrac{5}{2} \pi^2 r^3,$$

or the whole volume $= 5 \pi^2 r^3$.   See Ex. 15, Art. 147.

6. Find the volume generated by the revolution of the witch about its asymptote. $x^2 y = 4 r^2 (2\,r - y)$;

$$\pi \int y^2 dx = \pi \int \frac{64\,r^6}{(x^2 + 4\,r^2)^2} dx = 64\,r^6 \pi \int \frac{dx}{(x^2 + 4\,r^2)^2}$$

$$= 64\,r^6 \pi \left(\frac{x}{8\,r^2 (4\,r^2 + x^2)} + \frac{1}{16\,r^3} \tan^{-1} \frac{x}{2\,r} \right)$$

$$+ C \Big]_{-\infty}^{+\infty} = 4 \pi^2 r^3.$$

See Ex. 12, Art. 147.

7. Show that the volume generated by the revolution of $x^{\frac{2}{3}} + y^{\frac{2}{3}} = a^{\frac{2}{3}}$ about the axis of X is $\tfrac{32}{105} \pi a^3$.

8. Find the surface of the paraboloid of revolution.

$$S = 2\pi \int y\sqrt{dx^2 + dy^2} = 2\pi \int y\sqrt{\frac{y^2}{p^2} + 1}\,dy = \frac{2}{3}\frac{\pi}{p}(y^2 + p^2)^{\frac{3}{2}} + C.$$

Estimating the surface from the vertex, $S = 0$ when $y = 0$;

whence $C = -\frac{2}{3}\frac{\pi}{p}p^3$, and $S = \frac{2}{3}\frac{\pi}{p}[(y^2 + p^2)^{\frac{3}{2}} - p^3].$

9. Find the surface of the sphere.

$$S = 2\pi \int_{-r}^{r} y\sqrt{\frac{x^2}{y^2} + 1}\,dx = 2\pi \int_{-r}^{r} y\sqrt{\frac{y^2 + x^2}{y^2}}\,dx = 2\pi \int_{r}^{r} rdx = 4\pi r^2.$$

10. Find the surface generated by the revolution of $x^{\frac{2}{3}} + y^{\frac{2}{3}} = a^{\frac{2}{3}}$ about X.   *Ans.* $\frac{12}{5}\pi a^2$.

11. Find the surface generated by the revolution of the cycloid about its base.

$$S = 2\pi \int_{0}^{2r} y\sqrt{\frac{y^2}{2ry - y^2} + 1}\,dy = 2\pi\sqrt{2r} \int_{0}^{2r} y(2r - y)^{-\frac{1}{2}}dy$$

$$= -2\pi\sqrt{2r}(\tfrac{2}{3}(4r + y)(2r - y)^{\frac{1}{2}})]_{0}^{2r} = \tfrac{32}{3}\pi r^2.$$

Hence the whole surface $= \frac{64}{3}\pi r^2$. See Ex. 5, 143.

12. Prove that the surface generated by the revolution of one branch of the tractrix about X is $2\pi a^2$. See Ex. 5, Art. 158.

13. Prove the area of the surface of the prolate spheroid is

$$2\pi b^2 + \frac{2\pi ab}{e}\sin^{-1}e.$$

Analytic Geometry.

By A. S. HARDY, Ph.D., Professor of Mathematics in Dartmouth College, and author of *Elements of Quaternions.* 8vo. Cloth. xiv + 239 pages. Mailing Price, $1.60; for introduction, $1.50.

THIS work is designed for the student, not for the teacher. Particular attention has been given to those fundamental conceptions and processes which, in the author's experience, have been found to be sources of difficulty to the student in acquiring a grasp of the subject as a *method of research.* The limits of the work are fixed by the time usually devoted to Analytic Geometry in our college courses by those who are not to make a special study in mathematics. It is hoped that it will prove to be a *text-book* which the teacher will wish to use in his class-room, rather than a *book of reference* to be placed on his study shelf.

Oren Root, *Professor of Mathematics, Hamilton College:* It meets quite fully my notion of a text for our classes. I have hesitated somewhat about introducing a generalized discussion of the conic in required work. I have, however, read Mr. Hardy's discussion carefully twice; and it seems to me that a student who can get the subject at all can get that. It is my present purpose to use the work next year.

John E. Clark, *Professor of Mathematics, Sheffield Scientific School of Yale College:* I need not hesitate to say, after even a cursory examination, that it seems to me a very attractive book, as I anticipated it would be. It has evidently been prepared with real insight alike into the nature of the subject and the difficulties of beginners, and a very thoughtful regard to both; and I think its aims and characteristic features will meet with high approval. While leading the student to the usual useful results, the author happily takes especial pains to acquaint him with the character and spirit of analytical methods, and, so far as practicable, to help him acquire skill in using them.

John R. French, *Dean of College of Liberal Arts, Syracuse University:* It is a very excellent work, and well adapted to use in the recitation room.

Elements of Quaternions.

By A. S. HARDY, Ph.D., Professor of Mathematics, Dartmouth College. *Second edition, revised.* Crown 8vo. Cloth. viii + 234 pages. Mailing Price, $2.15; Introduction, $2.00.

THE chief aim has been to meet the wants of beginners in the class-room, and it is believed that this work will be found superior in fitness for beginners in practical compass, in explanations and applications, and in adaptation to the methods of instruction common in this country.

Elementary Co-ordinate Geometry.

By W. B. Smith, Professor of Physics, Missouri State University. 12mo. Cloth. 312 pages. Mailing Price, $2.15; for introduction, $2.00.

WHILE in the study of Analytic Geometry either gain of knowledge or culture of mind may be sought, the latter object alone can justify placing it in a college curriculum. Yet the subject may be so pursued as to be of no great educational value. Mere calculation, or the solution of problems by algebraic processes, is a very inferior discipline of reason. Even geometry is not the best discipline. In all thinking, the real difficulty lies in forming clear notions of things. In doing this all the higher faculties are brought into play. It is this formation of concepts, therefore, that is the essential part of mental training. And it is in line with this idea that the present treatise has been composed. Professors of mathematics speak of it as the most exhaustive work on the subject yet issued in America; and in colleges where an easier text-book is required for the regular course, this will be found of great value for post-graduate study.

Wm. G. Peck, *Prof. of Mathematics and Astronomy, Columbia College:* I have read Dr. Smith's Coordinate Geometry from beginning to end with unflagging interest. Its well compacted pages contain an immense amount of matter, most admirably arranged. It is an excellent book, and the author is entitled to the thanks of every lover of mathematical science for this valuable contribution to its literature. I shall recommend its adoption as a text-book in our graduate course.

Elements of the Theory of the Newtonian Potential Function.

By B. O. Peirce, Assistant Professor of Mathematics and Physics, Harvard University. 12mo. Cloth. 154 pages. Mailing Price, $1.60; for introduction, $1.50.

THIS book was written for the use of Electrical Engineers and students of Mathematical Physics because there was in English no mathematical treatment of the Theory of the Newtonian Potential Function in sufficiently simple form. It gives as briefly as is consistent with clearness so much of that theory as is needed before the study of standard works on Physics can be taken up with advantage. In the second edition a brief treatment of Electro-kinematics and a large number of problems have been added.

Academic Trigonometry : *Plane and Spherical.*

By T. M. BLAKSLEE, Ph.D. (Yale), Professor of Mathematics in the University of Des Moines. 12mo. Paper. 33 pages. Mailing Price, 20 cents; for introduction, 15 cents.

THE Plane and Spherical portions are arranged on opposite pages. The memory is aided by analogies, and it is believed that the entire subject can be mastered in less time than is usually given to Plane Trigonometry alone, as the work contains but 29 pages of text. The Plane portion is compact, and complete in itself.

Examples of Differential Equations.

By GEORGE A. OSBORNE. Professor of Mathematics in the Massachusetts Institute of Technology, Boston. 12mo. Cloth. vii + 50 pages. Mailing Price, 60 cents; for introduction, 50 cents.

A SERIES of nearly three hundred examples with answers, systematically arranged and grouped under the different cases, and accompanied by concise rules for the solution of each case.

Selden J. Coffin, *lately Prof. of* | Its appearance is most timely, and it
Mathematics, Lafayette College : | supplies a manifest want.

Determinants.

The Theory of Determinants: an Elementary Treatise. By PAUL H. HANUS. B.S., recently Professor of Mathematics in the University of Colorado, now Principal of West High School, Denver, Col. 8vo. Cloth. viii + 217 pages. Mailing Price, $1.90; for introduction, $1.80.

THIS book is written especially for those who have had no previous knowledge of the subject, and is therefore adapted to self-instruction as well as to the needs of the class-room. The subject is at first presented in a very simple manner. As the reader advances, less and less attention is given to details. Throughout the entire work it is the constant aim to arouse and enliven the reader's interest, by first showing how the various concepts have arisen naturally, and by giving such applications as can be presented without exceeding the limits of the treatise.

William G. Peck, *Prof. of Mathe-* | T. W. Wright, *Prof. of Mathemat-*
matics, Columbia College, N.Y.: A | *ics, Union Univ., Schenectady, N.Y.:*
hasty glance convinces me that it is | It fills admirably a vacancy in our
an improvement on Muir. | mathematical literature, and is a
(*Aug.* 30, 1886.) | very welcome addition indeed.

Mathematics.[2]

Copies sent to Teachers for examination, with a view to Introduction, on receipt of Introduction Price.

GINN & COMPANY, Publishers.

BOSTON.   NEW YORK.   CHICAGO.